WORLD BANK TECHNICAL PAPER NO. 493

District Energy Trends, Issues, and Opportunities

The Role of the World Bank

Carolyn Gochenour

The World Bank
Washington, D.C.

Copyright © 2001
The International Bank for Reconstruction
and Development/THE WORLD BANK
1818 H Street, N.W.
Washington, D.C. 20433, U.S.A.

ISBN: 0-8213-4911-2
ISSN: 0253-7494

Carolyn Gochenour is a principal financial analyst in the Energy Sector Unit of the Europe and Central Asia division at the World Bank.

Library of Congress Cataloging-in-Publication Data

Gochenour, Carolyn, 1949-
   District energy trends, issues, and opportunities: the role of the World Bank / Carolyn Gochenour
      p.   cm. — (World Bank technical paper ; no. 493)
   Includes bibliographical references
   ISBN 0-8213-4911-2
   1. Heating from central stations.   2. Air conditioning from central stations.   3. Heating and ventilation industry—Europe, Eastern—Finance.   4. Heating and ventilation industry—Europe, Central—Finance. 6. World Bank—Customer services. I. Title.

TH7641.G58   2001
   333.793—dc21   2001016271
                                                                  CIP

DISTRICT ENERGY TRENDS, ISSUES, AND OPPORTUNITIES

The Role of the World Bank

Annexes

FOREWORD

District heating (DH) is the dominant technology for the supply of heat and hot water in Eastern and Central Europe and the former Soviet Union (FSU) and can be expected to continue to operate in many places in these countries as the least-cost technical heating alternative for the foreseeable future. These heating systems, which are outdated and poorly maintained, represent some of the largest sources of energy wastage in Eastern and Central Europe and the FSU. As has been shown in a number of early investment projects supported by the World Bank, substantial energy efficiency savings can be obtained through rehabilitation and modernization of DH systems. These savings can help to reduce the cost of the DH service and improve its reliability and affordability. However, due to the limited investment funds available so far and the poor creditworthiness of municipalities, the main owners of DH systems, little progress has been achieved in improving the efficiency and functioning of DH systems in Eastern and Central Europe and the FSU.

Eastern and Central European and FSU countries will, for decades to come, compete with a growing number of other emerging markets for finance for their huge unmet investment needs. The policies and practices adopted by DH industry managers and regulators in these countries will greatly influence the flows of private capital to this sector. Experience from other parts of the world where DH systems exist could be beneficial to DH industry managers and regulators in Eastern and Central Europe and the FSU in modernizing their management systems and improving their policies and practices to allow for the attraction of capital that is urgently needed.

This study was, therefore, undertaken in order to record the worldwide DH and district cooling (DC) industry and trends and to discuss the key issues in DH in Eastern and Central European and FSU countries in order to arrive at operationally relevant issues and to define opportunities for new activities to be supported by the World Bank. The initiative is intended to increase awareness and understanding of the nature of interventions which can lead to increasing energy efficiency and the attractiveness of the sector to private financing.

Hossein Razavi
Director
Energy Sector Unit
Europe and Central Asia Region

ABSTRACT

This study reviews the history and development of district heating and district cooling, together referred to as district energy, in North America, Western Europe, Asia, Eastern and Central Europe and the Former Soviet Union (FSU). Experience from other parts of the world where DH systems exist could be beneficial to DH industry managers and regulators in Eastern and Central Europe and FSU countries in modernizing their management systems and improving their policies and practices to allow for the attraction of capital that is urgently needed. The report reviews, in a systematic manner, the key institutional, economic, financial, technical and environmental issues in the countries of Eastern and Central Europe and the FSU with the goal of arriving at operationally relevant issues and defining opportunities for new activities to be supported by the World Bank. The initiative is intended to increase the awareness and understanding of the nature of activities which can lead to increasing energy efficiency and the attractiveness of district energy to private financing.

ACKNOWLEDGMENTS

This report was prepared by Ms. Carolyn Gochenour, Energy Sector Unit, Europe and Central Asia Region, World Bank, largely on the basis of research carried out by Mr. Ishai Oliker, President, Joseph Technologies Corporation, New Jersey, United States and Ms. Lea Gynther, District Heating Specialist, Ekono Energy, Finland. Joseph Technologies Corporation and Ekono Energy are two of the leading district energy management consulting, design and engineering firms in the world. The report also benefited from advice given by Mr. Pentti Aro and Mr. Arto Nuorkivi, District Heating Engineers, Mr. Rachid Benmessaoud, Energy Planner, who is managing the Bank's first district heating projects in Poland, and Mr. Laszlo Lovei, Ms. Anke Meyer and Mr. Kyran O'Sullivan, Energy Specialists of the World Bank, among a number of other contributors. The study incorporates the findings of the World Bank's early interventions in district heating in Poland, Estonia, Latvia, Ukraine and China.

ACRONYMS AND ABBREVIATIONS

AIJ	Activities Implemented Jointly
BAT	Best Available Technology
CAA	Clean Air Act
CHP	Combined-Heat-and-Power
CDM	Clean Development Mechanism
CFC	Chloro-Flouro-Carbon
CH_4	Methane
CMEA	Council for Economic Assistance
CO_2	Carbon Dioxide
COP	Coefficient of Performance
DC	District Cooling
DE	District Energy
DH	District Heating
DHE	District Heating Enterprise
DWSC	Deep Water Source Cooling
EBRD	European Bank for Reconstruction and Development
ECA	Europe and Central Asia
EIB	European Investment Bank
EMAS	Eco Management and Audit Scheme
EPA	Environmental Protection Agency
ERs	Emission Reductions
ESCO	Energy Service Company
EU	European Union
FRP	Fiber-reinforced polymer
FSU	Former Soviet Union
GDP	Gross Domestic Product
GEF	Global Environment Facility
HCFC	Hydrogenated Chloro-Flouro-Carbon
HCl	Hydrochloric Acid
HFC	Hydro-Flouro-Carbon
HOB	Heat-only-Boiler
ICAP	Installed Capacity
ISO	Independent System Operator or International Standardization Organization
JI	Joint Implementation
MDB	Multilateral Development Bank
NUTEK	Swedish Aid Agency for New Technology
NIB	Nordic Investment Bank
NO_x	Nitric Oxide
N_2O	Nitrous Oxide
NUG	Non-utility Generator
O&M	Operations and Maintenance
PCF	Prototype Carbon Fund
PSC	Public Service Commission
PURPA	Public Utilities Regulatory Policy Act
SAL	Structural Adjustment Loan
SIL	Sector Investment Loan
SO_x	Sulfur Oxide
UNFCCC	United Nations Framework Convention on Climate Change
US	United States
VAT	Value-Added Tax
VOC	Volatile Organic Compound
WHO	World Health Organization

WEIGHTS AND MEASURES

Btu	British Thermal Unit
Gcal	Gigacalorie (10^9 calories)
GJ	Gigajoule (10^9 joules)
GW_t	Gigawatt thermal (10^9 watts)
GWh	Gigawatt-hour (10^9 watt-hours)
kg	Kilogram (10^3 grams)
km	Kilometer (10^3 meters)
kWh	Kilowatt-hour (10^3 watt-hours)
m^2	Square meter
MPa	Megapascale
MW_c	Megawatt cooling (10^6 watts)
MW_e	Megawatt electricity (10^6 watts)
MW_t	Megawatt thermal (10^6 watts)
MWh	Megawatt-hour (10^6 watt-hours)
PJ	Petajoule (10^{15} joules)
TJ	Terajoule (10^{12} joules)
TWh	Terawatt-hour (10^{12} watt-hours)

CONVERSION FACTORS

1 Gcal = 4.187 GJ = 1,163 kWh

1 ton refrigeration = 12,000 Btu = 12,648 kJ

Executive Summary

Introduction

This study reviews the history and development of district heating (DH) and district cooling (DC), together referred to as district energy (DE), in North America, Western Europe, Asia, Eastern and Central Europe and the Former Soviet Union (FSU). The report provides a systematic review of the worldwide practices and latest state-of-the-art in technical developments, institutional and regulatory arrangements, economic justification, financial arrangements and environmental thinking in order to provide insights and guidance to addressing the issues in the Bank's client countries, primarily those in Eastern and Central Europe and the FSU. The initiative is intended to increase the awareness and understanding of the nature of activities which can lead to increasing energy efficiency and the attractiveness of district energy to private financing.

Extent of Coverage of DE Systems

DH systems exist in the United States and Canada in North America and also in many Western European, Eastern and Central European and FSU countries as well as in Asia in Japan, Korea, China and Mongolia. DH systems were introduced in the beginning of the twentieth century, with rapid development after World War II. Today, the number of DH consumers in Europe is over 100 million. Over 50% of Europe's total DH consumption is in Russia, with a further 26% concentrated in the neighboring Eastern European countries. Western Europe accounts for 20%, of which 40% is concentrated in the Nordic countries. DC systems have been introduced more recently in Europe. In the United States, DE systems are less extensively utilized and provide only about 4% of space heating and cooling demand.

Competitive Advantages of DH and DC

DH offers a number of benefits over decentralized heating options in areas of high heat load density. One of the key features improving the competitiveness of DH has been the increased use of combined-heat-and-power production (CHP), i.e. co-generation, which increases the efficiency of primary fuels' use by 35-40% compared to condensing power production and heat-only-boilers (HOBs). The use of primary energy (fuels) has been significantly reduced due to CHP production and DH use. This, in turn, has in many sites had a significant positive environmental impact on the ambient air quality. Other key characteristics in favor of DH are the wider variety of fuels that can be used in the case of DH as compared to building-level boilers - coal, oil, natural gas, refuse-driven fuels, geothermal, peat and biomass– and the flexibility DH production plants provide for switching into another fuel. DC offers the same benefits as DH and can be integrated with DH systems to further reduce use of primary fuels.

However, in areas of low heat low density, milder climates or where the value of fuel savings in CHP or HOB plants, as compared to individual building-level boilers, is low, DH may not be the least-cost option. In such cases, it may be more advantageous to establish new, decentralized heating systems and abandon DH.

Key Characteristics of DH Systems in the West and in Eastern and Central Europe and the FSU

The technical features of Western European systems, based predominately on a distribution medium utilizing hot water, are of a high level of efficiency, whereas the DH systems in North America, based on a distribution medium utilizing steam, are less efficient. Eastern European systems were mainly established during the influence of the former Soviet Union and are based on technology developed in the Soviet

Union. The main technical differences of Eastern and Western European systems are found in their technical configurations, performance and efficiency. In general, Eastern European systems are in poor condition, with much lower efficiencies and higher heat and water losses, and are therefore in need of renovation and reconstruction.

Both Eastern and Western European DH utilities are characterized by predominately public ownership, although privatization of DH utilities, mainly CHP plants, is increasing. In the Western European energy business, DH and electricity activities are mostly concentrated in the same company, while in Eastern Europe, these activities are normally divided into two or more different companies. In order to best optimize the DH operations of a system within a city so that DH can compete with other heating alternatives, a number of Eastern European countries have started to merge the various DH companies serving the same area so that heat production, transmission and distribution are contained in the same company, as was the model in the West from the beginning.

While DH is typically viewed as a "natural monopoly," it has, in principle, always been competing with other energy alternatives, such as individual boilers fired by oil or solid fuels and fixed networks supplying natural gas and electricity. As such, it has usually been brought under the same kind of regulatory control as other utilities. However, in Scandinavia and some other parts of Europe, DH is not subject to regulation by regulatory authorities, as it is judged to be regulated by the market, since it is operating in an inherently competitive environment.

In Eastern Europe, the allocation of expenses in the joint production of heat and electricity in CHP plants has generally resulted in the benefits of the joint production being allocated to electricity rather than attempting to share the benefits with the two products. This results in prices for heat from the CHP plants to be at the same level as heat produced in HOBs or even higher. The deficiencies of this pricing method are being recognized in a number of Eastern European countries, especially as consumers now have other alternatives to DH.

In Western Europe, DH utilities usually combine the market-oriented and cost-based approaches to tariff-setting, establishing tariffs close to, but lower than, the next alternative cost of heat supply. DH tariffs reflect the actual costs of serving different consumer categories, and the tariffs encourage energy conservation and are easy to understand. In Eastern Europe, on the other hand, tariffs are usually based on estimated consumption and generally do not reflect the actual costs of supply. In addition, oftentimes, tariffs are established below cost recovery levels, thereby requiring subsidies from their owners, usually the municipalities.

Billing and collection systems also differ significantly in Western and Eastern Europe. In Western Europe, each house, apartment building, industry and other consumer is individually metered and payment is often carried out by direct debit of the consumer's bank account by the DH utility, resulting in good payment performance. In Eastern Europe, however, only large consumers are typically metered and payment performance of non-metered consumers is generally poor. In fact, the main financial problem facing DH utilities in Eastern Europe is the high level of unpaid heat and hot water bills.

A typical feature of billing in FSU countries is the practice by DH utilities of contracting with intermediaries known as municipal house maintenance companies to prepare bills and collect payments for heat and hot water services, along with the payments of other utility bills, from residential consumers. Since DH utilities do not maintain the individual payment records of residential consumers, the utilities are not in a position to identify which residential consumers have not paid and therefore cannot disconnect consumers. Many DH utilities in Eastern Europe and the FSU are initiating schemes to introduce direct billing to residential consumers after individual building-level meters are installed, which then allows for disconnection of non-paying consumers.

In many Eastern European countries and the FSU, social protection is provided for low-income households in order to mitigate adverse social impacts from increasing prices for district heating and hot water services. Household surveys in Ukraine, Latvia and Russia have shown that targeted housing allowance programs have serious flaws, typically with insufficient coverage of deserving households and substantial leakage to non-eligible households. Improvements in means testing and administration of social assistance programs is needed to enhance their effectiveness.

The options for financing of needed investments is more limited in Eastern European countries than in Western European countries. Typically Western European DH utilities finance required investments in their systems through self-financing methods and borrowed capital. Eastern European DH utilities, on the other hand, find that one of their major problems is to secure financing for needed investments and to replace the traditional sources of capital funds which were provided from state or local budgets. Today, the most relevant financing instruments available include loans and equity investments of multilateral development banks and suppliers' credits.

The Future Potential of DH and DC

The expectations regarding the development of market share of DH vary significantly country by country. Throughout Eastern Europe, the competition between different heating forms is becoming more intense as the energy markets are being liberalized. In some countries, DH has been extended beyond what is feasible and the market share may therefore decrease. Also, the specific heat consumption of buildings may improve significantly as measures are implemented to improve the energy efficiency of buildings, which may reduce the total heat demand and even the viability of DH in some areas.

On the other hand, in some European countries, the market share of DH is expected to grow due to increased use of CHP and increased emphasis on environmental considerations, especially climate change. The EU energy strategy emphasizes the increased use of CHP for which extension of DH use is a prerequisite. New technologies and technical improvements will make DH systems more efficient.

In some Western and Eastern European countries, the political decisions on closing nuclear power plants, if implemented, will have a significant impact on the demand for new CHP capacity construction. In line with power market liberalization, as old power plants in Europe are retired, the new power plants are most likely to be CHPs, which are more competitive than alone-standing condensing power plants. The differences in the development potential of new CHP plants between the countries in Europe are large and depend on the available heat markets.

In addition, air quality guidelines are becoming more stringent in Eastern Europe and will also have an important impact on the future development of DH. Today, ambient air quality does not meet the applicable guidelines in all countries. One of the most efficient ways to improve air quality quickly is to reduce the use of small boilers in buildings and to switch to DH.

Though the current use of DC is rather limited in Europe, the application of DC is increasing quite rapidly. The existing DH networks and CHP plants enable rather easy practical adaptation of DC technology. DC is also growing in the United States and Asia.

High Energy Wastage in DH Systems in the Bank's Client Countries

Today, in many countries of Eastern and Central Europe and the FSU, DH systems represent some of the largest sources of energy wastage due to years of lack of maintenance and technical upgrading. For a long time, low cost energy supplies from Russia allowed these countries to postpone the introduction of modern energy efficiency technologies in supply and demand as well as the process of commercialization and restructuring which was taking place in Western countries that were facing much higher international energy prices. Energy efficiency savings of more than 30% are typically feasible through rehabilitation or replacement of equipment and introduction of automation and control equipment. These savings could help to reduce the cost of the DH service, which is usually the single largest expenditure in a household's budget in Eastern and Central European and FSU countries, and to improve the balance of payments of energy importing countries. However, little progress has been achieved so far due to lack of funds and poor creditworthiness of municipalities, the main owners of DH systems.

Huge Investment Requirements in Eastern and Central Europe and the FSU

An estimate of the order of magnitude of investment needs in the Eastern and Central European and FSU countries has been roughly calculated, based on the requirements for investments which were needed to bring about at least a 20% energy savings in the Bank's recently completed projects in four Polish cities. A regression analysis was carried out utilizing the project costs and the heat sales of the individual Polish DH systems to compute a formula which could be applied to other countries where data on heat sales is available.

For 11 countries in Eastern and Central Europe and the FSU where heat sales data is available, including Russia, Ukraine, Romania, Poland, Czech Republic, Hungary Lithuania, Estonia, Bulgaria, Croatia and Slovenia, the total annual heat sales in 1999 was about 3,700 PJ and investment requirements, excluding those already undertaken, are estimated to be huge at about US$ 25 billion over a 5-7 year period. When the remaining 15 Eastern and Central European and FSU countries are considered, the total investment requirements would be considerably higher.

Limited Availability of Financing for DH Investments

Since DH companies in Eastern and Central Europe and the FSU have not been able to build up adequate reserves for future investments through their tariff policies, funds for investment requirements must be raised from external sources, such as loans or private equity. However, the possibility to raise funds from these sources at terms suitable for infrastructure investments is virtually non-existent in Eastern and Central Europe and the FSU. The local commercial banking sector in many countries is still underdeveloped, but even where local loans for investment purposes are available, they are usually only available at terms which do not match the requirements of typical DH projects. The experience to-date with raising equity from private investors is also very limited. So far, interest from private investors has been concentrated on CHP plants, which are typically separate enterprises from the DH network utilities, or on electricity enterprises which also include the DH business.

Sustainable private participation in infrastructure services requires certain conditions which are generally lacking in Eastern and Central European and FSU countries, both at the macro level and the sector level, particularly the potential to recover costs and to exert leverage over customers to encourage payment discipline. The inability of most DH enterprises to disconnect customers because of the lack of metering, the use of intermediaries in billing and the lack of a direct relationship with the customers leaves DH enterprises little leverage over customers to encourage payment. In this sense, the DH sector differs significantly from the electricity sector and may help to explain why private investors who are willing to enter countries in Eastern and Central Europe and the FSU in the electricity sector may find the DH sector unattractive. Since DH is not common throughout the world, the absence of a project demonstrating the privatization of DH services and the lack of industry specialists and financiers with privatization experience in the sector may also be reasons why private sector participation in DH is not a common initiative.

In the FSU and Eastern and Central European countries, financing from multilateral development banks is therefore likely to remain as the main financing instrument over the near term as a consequence of their still relatively less developed economies, poorly developed capital markets, lack of access to long-term financing and limited opportunities for attracting private investment.

A Special Role for the World Bank

The early DH rehabilitation projects completed or underway have begun to demonstrate the very positive results that can be obtained from Bank-supported interventions. During the transition, the World Bank can play an important role in assisting Eastern and Central European countries in a number of ways.

Development of Least-Cost, Long-Run Heating Strategies. The World Bank can assist, both at the country-level and at the city-level, to develop heating strategies, which take into account the factors necessary to determine the least-cost, long-run alternative(s) most appropriate for those areas. DH systems were the preferred and even mandatory approach to heating in Eastern Europe, but DH has not always been the least-cost solution. Master plan studies or country-wide strategy studies can be designed to look beyond current energy price levels and short-term distortions to determine the optimum long-run option for heating. The World Bank has now developed considerable expertise to be particularly effective to help guide such strategy studies in these areas.

Poverty Alleviation. The World Bank, through its heating investment projects, can also directly assist poor households connected to DH or utilizing other heating options in a number of ways. First, by participating in the design of projects which would reduce the cost and improve the quality of energy supplied, and thereby lower heat tariffs and improve affordability of heating services, the Bank is helping to improve and better ensure access to heating and hot water by the poor.

Secondly, through investment projects which improve heat service reliability and affordability, household expenditures can be lowered for supplemental heating, and resources by all households, including poor households, can be increased for other energy efficiency investments, such as weather-stripping of windows and doors inside apartments and insulation of roofs inside buildings, which further help to reduce heating bills.

Thirdly, by working together with central and local governments to improve the targeting of energy subsidies, the Bank is helping to better ensure that social assistance reaches the poor and vulnerable groups.

Fourthly, by promoting energy efficient and less polluting end-use technologies, the Bank is working to improve health conditions affecting both the poor and other households.

Supporting Macro-Fiscal Stabilization. World Bank-supported heating projects in Eastern and Central Europe can lead to significant macro and fiscal impacts. Fuel savings through efficiency improvements have now been demonstrated to typically amount to more than 25-30% in the early projects, and these savings translate into direct improvements in the balance of payments for fuel importing countries.

Heating-related expenditures are a high burden for many municipal budgets, and increasing efficiency and reducing costs of heat supply can help to reduce the burden on municipalities by providing financing for urgently-needed capital investments, reducing the level of operational subsides needed by heating utilities, and reducing the level of subsidies needed by low-income and vulnerable households.

Promoting Environmental Sustainability. The World Bank can also play a leading role in helping countries in Eastern and Central Europe to protect the environment through investment projects in DH, or in decentralized heating systems, where the scope for reducing harmful emissions is high. The improvements in efficiency in the early World Bank-supported DH and energy efficiency projects have resulted in significant reductions in fuel use, leading in turn to reductions of emissions, particularly CO_2 and other greenhouse gases, important in mitigating climate change.

The early projects have also helped to protect the environment by promoting the switch away from coal and heavy fuel oil, to cleaner fuels, such as gas or wood products, in heat production and by including measures to improve pollution controls.

A number of projects, some supported by the Global Environment Facility (GEF), have also been important in assisting to remove the barriers to renewables and energy efficiency investments, further helping to protect the environment. Through such investments, the World Bank is further assisting countries in Eastern and Central Europe, many of which are in the process of EU Accession, to move in line with EU policies promoting greater energy efficiency and environmental protection.

Furthermore, through its recently-established Prototype Carbon Fund, the World Bank will pilot the production of emissions reductions within the framework of Joint Implementation and the Clean Development Mechanism. Substantial opportunities for trading carbon credits exist in the area of DH and other types of heating investment projects in Eastern and Central European countries.

Improving Regulation of Heat Pricing. By applying its global experience in the heating sector, the World Bank is also helping countries to develop appropriate pricing policies and to establish a clear and independent regulatory framework to improve energy regulation. This is especially important in the areas of bulk heat pricing from CHP plants, where the benefits of the co-generation process are generally not shared with DH, and gas pricing, where lack of differentiation of prices among consumer groups, with small consumers typically paying the same as large consumers, creates unfair competition to DH. The World Bank has been proactive with its clients and the regulators in Eastern and Central Europe in working towards full cost recovery of heating services, the elimination of cross-subsidies in heat tariff structures and the introduction of two-tier heat tariffs to promote greater transparency in heat pricing. Clear and transparent regulation of heat tariffs, in turn, helps to reduce an important barrier to private sector involvement.

Demonstration Impacts with Spillover Effects. While the early World Bank projects in DH have been focused on particular localities, these early projects have had important demonstration effects in their respective countries, and this is already leading to further investments in other localities.

Catalyzing Additional Funding for Heating Investments. The World Bank-supported interventions in DH and energy efficiency are helping to mobilize co-financing from other international financial institutions and donor organizations. By taking the lead in helping to set the overall policy framework and build the institutions in the sector, the World Bank has been able to attract substantial additional financing from other loan and grant sources. In this way, the share of World Bank financing of investment projects can be reduced as other organizations take a larger role in financing.

Supporting Private Sector Development. World Bank-supported investment projects in DH and energy efficiency are starting to demonstrate the efficiency gains and institutional and financial improvements which can be achieved by DH enterprises, and other heating agencies, in countries in transition. The World Bank can thus be seen as paving the way for greater commercial financing and private sector investments, as DH and other heating utilities are commercialized, which reduces the risks and makes the heating sector more attractive for private financing. The results of the recently-completed projects show that the private sector is willing to invest in DH, once the economies develop to a stage where adequate legal and regulatory frameworks are in place and the energy utilities have been commercialized and corporatized. Other projects are beginning to demonstrate that commercial banks may be willing to participate in financing investments at suitable terms when the World Bank is helping to establish the policies in the sector and the DH enterprise is operating on a commercial basis. These results are encouraging for fostering further private sector involvement in the future.

Conclusion

As this report has shown, the World Bank is well positioned to respond to the opportunity and challenges posed in the energy sectors in Eastern and Central Europe and the FSU today in a number of ways. The World Bank's involvement in a variety of early DH and energy efficiency projects has clearly demonstrated the Bank's special role in these fields. These early interventions have taken place at a time when these countries were facing severe resource constraints and lacked access to private capital at appropriate terms. Based on the early results, the World Bank has shown that it can play an important role during the transition period to market economies by providing capital to support the much-needed policy and institutional changes and heating sector investments until the sector is able to attract a sufficient volume of capital from other sources.

1. Introduction

A. Background

Some of the largest sources of energy wastage in Eastern and Central Europe (ECA) and the Former Soviet Union (FSU) are the outdated and badly maintained district heating (DH) systems. Modernization of heat generation facilities, DH transmission and distribution systems and consumer installations is of critical importance. Energy efficiency savings of more than 30% are typically feasible through rehabilitation or replacement of equipment and introduction of automation and control equipment. These savings could help to reduce the cost of the DH service, which is usually the single largest expenditure in a household's budget in ECA and FSU countries. However, little progress has been achieved so far due to lack of funds and poor creditworthiness of municipalities, the main owners of DH systems.

ECA and FSU countries have huge unmet investment needs. For decades to come, they will compete with a growing number of other emerging markets for finance. The policies and practices adopted by DH industry managers and regulators in these countries will greatly influence the flows of private capital to this sector. Experience from other parts of the world where DH systems exist could be beneficial to DH industry managers and regulators in ECA and FSU countries in modernizing their management systems and improving their policies and practices to allow for the attraction of capital that is urgently needed.

The World Bank has recently prepared a study on DH in ECA and FSU, under funding from ESMAP. The ESMAP study focused its attention on the factors which determine the choice of the economically preferred heating option from a set of alternatives and the circumstances under which DH or other options would be preferred. For the purpose of that study, a number of cases studies were examined utilizing the same methodology to determine the most cost-effective technology for the supply of heat. In the majority of cases, DH was shown to be the dominant technology for the supply of heat. It can be anticipated that DH systems will continue to operate in ECA and FSU countries as the least-cost technical heating alternative for the foreseeable future.

B. Objectives

The objective of this report is to document the worldwide DH and district cooling (DC) industry and trends and to discuss the key institutional, economic, financial, technical and environmental issues in DH in ECA and FSU countries in a systematic manner with the goal of arriving at operationally relevant issues and defining opportunities for new activities to be supported by the World Bank.

The initiative is intended to increase the awareness and understanding both within the World Bank and in the World Bank's client countries of the nature of interventions which can lead to increasing energy efficiency and the attractiveness of the sector to private financing. The activity would build upon the findings of the ESMAP-sponsored case studies on DH.

C. What Are District Heating and Cooling?

A district heating system is a concept in which heat is produced centrally in precise location(s), from where heat is distributed to the consumers located in different buildings, in the form of hot water or steam circulating in a distribution piping network. Often, heat is also used not only to heat buildings but also to provide domestic hot water and for industrial purposes, such as process heat.

A district cooling system is a concept in which the production of cooling is centralized in precise location(s), from where cold is distributed to the consumers located in different buildings, in the form of chilled water circulating in a distribution piping network. Alternatively, cold can be produced from heated water circulating in the network by using absorption technology, which enables the co-existence of DH and DC. Together, DH and DC are referred to as district energy (DE).

1

2. Extent and Key Characteristics of District Heating Systems

A. Early Development of District Heating Systems

The idea of producing and distributing heat in the form of hot water and steam can be traced back to 1622 when the Dutchman Cornelius Drebbel proposed a DH system based on a model of fresh water supply. Some historians say that DH existed in England in as far back as 1777. The history of DH in Russia dates back to the year 1832 when the first gravity heating system was commissioned.

In the United States, the beginning of DH can be traced back to the late eighteenth century when Benjamin Franklin sold heat to several adjacent residences in Philadelphia. Almost one hundred years later in 1877, Birdsill Holley designed the first financially successful DH system in Lockport, New York. This system, based on the delivery of steam, was widely imitated. By 1887, twenty DH systems were in operation in the United States; co-generation and DC were introduced as early as 1890.

After these early attempts, wider introduction of the systems began in the beginning of the twentieth century. The purpose was to rationalize ways of heating a number of buildings from one boiler through a suitable distribution medium. In the United States, the distribution medium was steam, while in Europe, the predominant distribution medium was hot water. After applications in heating of buildings used for apartments, offices and commerce, DH was also used for industrial applications.

In Europe, before World War II, the countries in the forefront in the development of DE systems were Germany, Denmark and Russia. In Germany, the first DH transmission pipeline was installed in Hamburg in 1893 to supply heat to the city hall. In Denmark, the first DH utility started its operations in Frederiksberg in 1903 by producing heat in a waste incineration plant and supplying it to a new hospital and other buildings. The second was in Copenhagen in 1925. In Russia, the first major system was introduced in St. Petersburg (Leningrad at the time) in 1924, and the second in Moscow in 1928. Numerous other schemes were installed in these three countries by the time of World War II. DH systems were introduced also in smaller numbers in Poland (1907), United Kingdom (1911), Czechoslovakia (1922), the Netherlands (1923), Paris, France (1928), Iceland (1930) and Switzerland (1934). Figure 2-1 summarizes the early history of DH, showing the earliest installations in a number of countries.

The potential for using waste heat from electricity production was recognized at an early stage. Waste heat was used in some places, and, as time went by, combined-heat-and-power (CHP) became more widely adopted. Today, CHP is acknowledged as a very effective way of producing electricity and heat. Rapid development of DH systems in Europe began after World War II. Today, the number of DH consumers in Europe is over 100 million. Over 50% of the Europe's total DH consumption is in Russia. A further 26% is concentrated in the neighboring Eastern European countries. Western Europe accounts for 20%, of which 40% is concentrated in the Nordic Countries.[1]

After World War II, about 250 urban steam DE systems were in operation in the United States serving a broad base of residential, commercial and industrial customers.

DH/DC systems were installed more recently in Asia where they can be found in Japan, the Peoples Republic of China, South Korea and Mongolia.

As DH is a necessity and DC is not, DH/DC systems did not develop in the countries of Latin America and Africa because most of these countries have a warm climate and are too poor to afford DC. In some

---

[1] Utility Europe 7/98, Euroheat & Power, 1999.

locations in South America where the climate is as cold as in the southern part of Eastern Europe, topography may have been a limiting factor to the possibilities to use DH.

FIGURE 2-1

Early History of District Heating

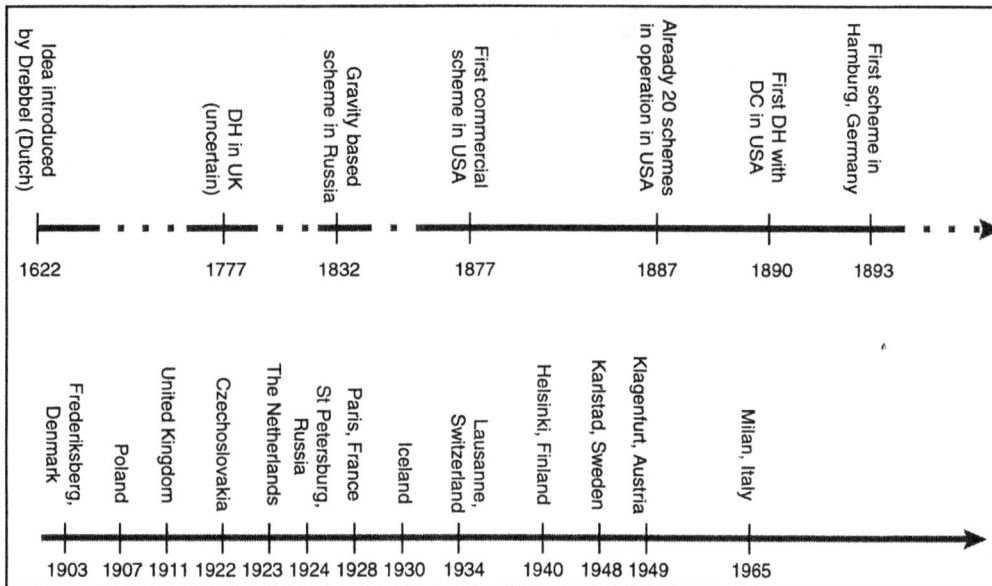

B. Western European District Heating Systems

Extent of Coverage

In 1991 the market share of DH in the space heating market of Western Europe was 7%.[2] In 1997, the leading users of DH (over 40% market share of heat use) were the Nordic countries of Iceland, Denmark, Finland and Sweden as shown in Figure 2-2. The next group of substantial users include Austria and Germany with a market share of about 12% (10% in 1995). Minor users include France, Netherlands, Norway, United Kingdom, Switzerland, Italy, Greece and Portugal. However, in spite of the low market share of DH in some countries (e.g. Germany and France), the total consumption may be as high or higher than in countries with higher market penetration (e.g. Finland).

Three of the thirteen largest single schemes in the world are installed in the Western Europe in Berlin, Copenhagen and Helsinki. Two of them are in the capitals of North European countries with rather harsh climatic conditions: Copenhagen (325,000 households, 7 000 GWh/year) and Helsinki (90% of buildings, 6,500 GWh/year). The system in Berlin supplies 11,000 GWh/year.[3] Other major Western European users are Stockholm (majority of the 800,000 inhabitants) and Vienna (over 2 million inhabitants including suburbs). However, DH can be found in many other Western European cities where the heating season is much shorter, such as Paris (France) and Brescia, Torino and Verona (Italy). Modern DH systems in Western Europe deliver heat and hot water year round.

---

[2] Hasenkopf, 1995.
[3] Koljonen, 1988.

3

FIGURE 2-2

Share of DH Households in Western Europe in 1997

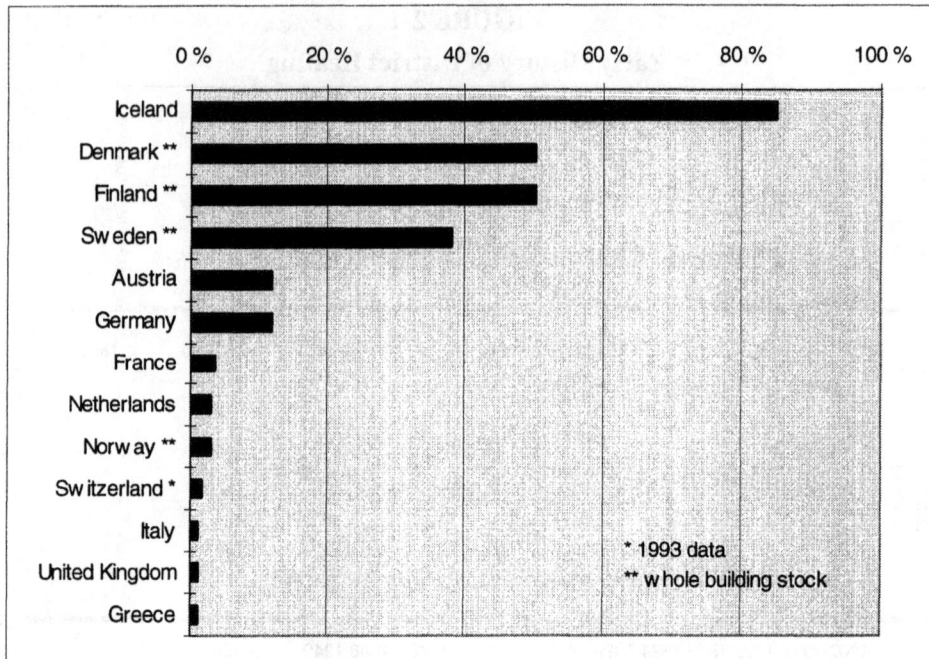

The annual energy consumption and length of pipeline systems in some Western European countries are shown in Figure 2-3 below. The figure portrays the differences in the sizes of the schemes in terms of energy carried and length of installations. Also, energy consumption per kilometer varies significantly among the different countries.

FIGURE 2-3

Annual Energy Consumption (GWh/year) and Length of Pipelines (km) in Some Western European Countries in 1996-97

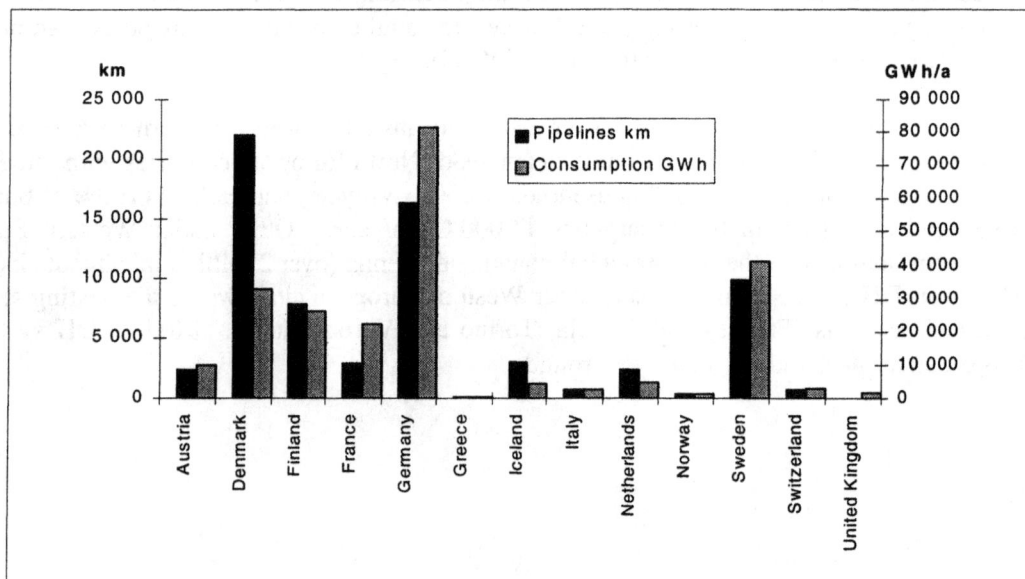

4

Co-generation

DH has, in principle, always been competing with other energy alternatives, such as individual boilers fired by oil or solid fuels and fixed networks supplying natural gas and electricity. One of the key features improving the competitiveness of DH has been the increased use of CHP production, i.e. co-generation, which increases the efficiency of primary fuels' use by 30-40% compared to condensing power production and heat-only-boilers (HOBs). This, in turn, has in many sites had a significant positive impact on the ambient air quality.

The share of DH produced by CHP plants varies country by country as shown in Figure 2-4. Also, the average size of installations varies from the typical under one megawatt in the United Kingdom to a few hundred megawatts in the Northern Europe. In most Central European countries, there is a growing trend in CHP use. In some other countries, such as Finland, the possibilities to increase the use of CHP are limited due to its currently high share both in absolute and relative terms.

FIGURE 2-4
Share of DH Produced by CHP in Western Europe in 1997

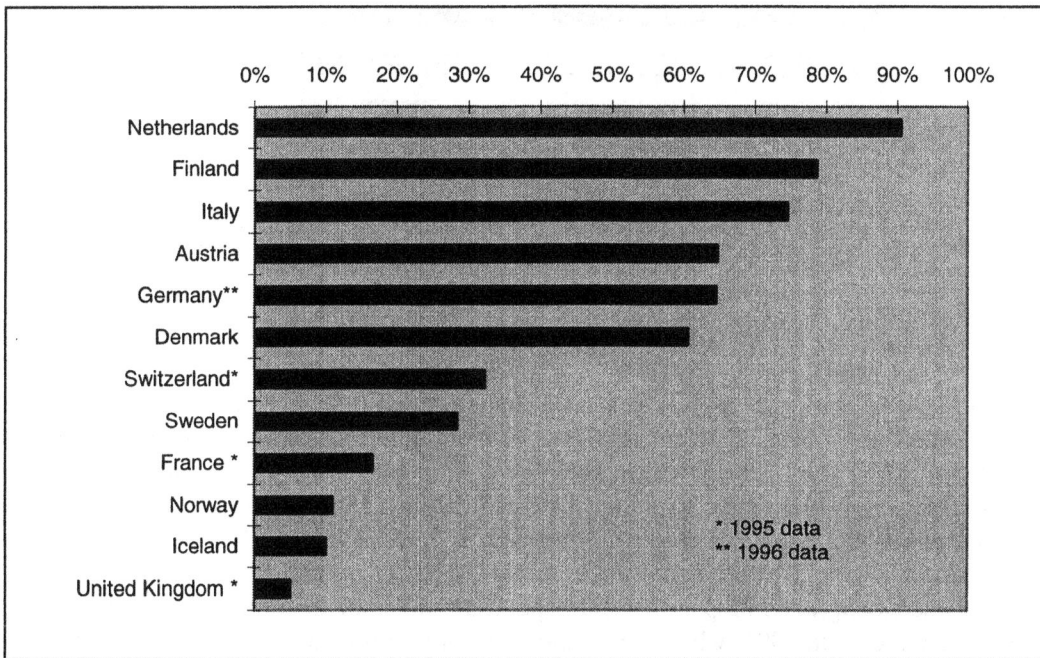

Fuels Utilized

One of the main characteristics in favor of DH is the variety of fuels that can be used for production: coal, oil, natural gas, refuse-derived fuels, peat and biomass. The production mix varies significantly country by country as shown in Figure 2-5. Austria, Italy, the Netherlands and United Kingdom rely heavily on natural gas whereas Germany, Finland and Denmark utilize substantial shares of coal. The main primary energy source for DH in Iceland is geothermal, and one of the important energy sources in Sweden is biomass. The share of natural gas has been increasing in a number of countries.

5

FIGURE 2-5

Fuels Used for DH Production in Western Europe in 1997

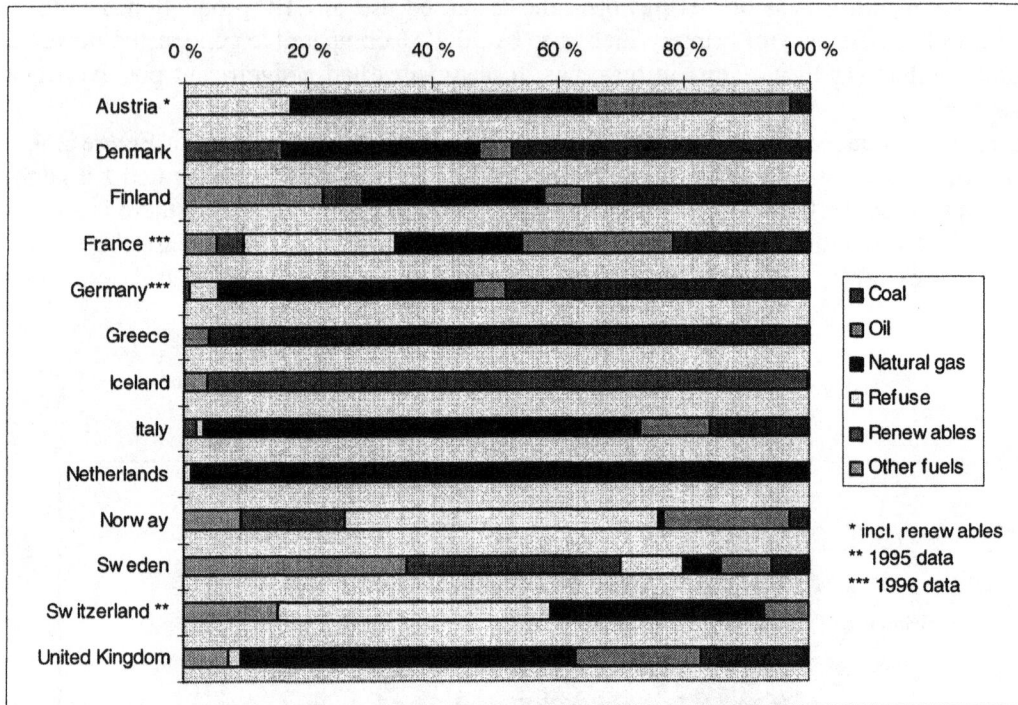

Technical Features

Heat Production. CHP plants are typically utilized in conjunction with DE systems in Western Europe. Technically, a CHP plant is a more efficient way to produce heat and power than separate power generation and heat production plant. Typical efficiencies of CHP plants are around 80-90% as compared to 35-45% in condensing power production and 90% for large HOBs. The use of primary energy (fuels) has been significantly reduced due to CHP production and DH use, especially in Denmark, Finland, Italy and Sweden. For example, with CHP production connected to DH, Denmark saves about 30 PJ of fuel annually. In Finland, the saving is 65 PJ per year corresponding to 5% of the total energy consumption. In Italy, the savings were 5.6 PJ in 1995, which represents 20% of fuel use for space heating. In Sweden, the annual saving is about 30 PJ corresponding to about 2% of the total national energy consumption.[4]

District Heating Networks. Preinsulated pipes have become the predominant piping technology in Western Europe. This technology reduces the level of corrosion in piping systems as compared to steel pipes laid in concrete channels. Typical heat losses in Western European DH networks are 4-10%, depending on the size and condition of the system. Water losses are typically small, with little need for make-up water. The state-of-the-art systems do not require more than one or two refillings per year.

The dominant mode of network operation in Western European systems has been based on variable flow regime. Load can be dispatched from several heat sources simultaneously to the same network, which has several benefits including optimization of the use of CHPs and HOBs, reduced pumping costs, increased reliability of heat supply, and reduced investments in transmission pipelines and pumping capacity.

---

[4] Euroheat & Power, 1997.

Consumer Installations. Consumer connections in Western Europe are usually indirect, which means that heat is transferred from the DH networks by means of heat exchangers to the internal heating networks of the buildings, thereby resulting in hydraulically separated, DH and building heating systems. However, variations exist among the countries. For example, direct connections, whereby water from the DH network is transferred directly to the internal heating network of the building and radiators, are used to some extent in Denmark and Germany, whereas only indirect connections are utilized in Finland. Consumers can regulate heat consumption, since every radiator is equipped with a valve, usually an electronic one.

Metering. Heat meters are typically installed in each house or in each apartment building. Heat meters are not installed in each individual apartment. In many countries, such as Germany and Austria, heat cost allocators, small devices installed on room radiators to allocate the building-level heating bill to the individual apartments, are used.

Operation and Maintenance. Western European maintenance procedures focus on preventive maintenance rather than on repair of damage that has already occurred. Preventive maintenance is carried out by regular monitoring programs of critical points in the DH system, such as chambers which are prone to flooding. Ensuring good drainage and ventilation of the channel type pipelines is a key element of good maintenance practice. Preventive maintenance is supported nowadays by modern computer-based operation and maintenance (O&M) systems, which facilitate maintenance work and enable follow-up reporting. The benefits are decreasing leaks and water losses and longer lifetimes of the technical assets of the DH systems. The reliability of heat supply is typically very good in Western European DH systems. For example, in Helsinki the average time a consumer suffered from interruptions due to repair of existing networks or construction of new networks was less than three hours in 1996.

The key characteristics of DH in the above Western European countries are described in more detail in Annex 1.

C. North American District Energy Systems

Extent of Coverage

District energy systems are considerably smaller, less extensively utilized and less efficient in the United States and Canada in North America than in Western Europe. In the United States, DE provides only 1.5% of total energy consumption and about 4% of the space heating and cooling demand. In particular, DH provides about 8% of the energy consumed in the commercial sector. As a result of its historical development, America has inherited a technology that is far from optimal in the form of the high pressure steam systems used in all older urban systems and the bulk of institutional systems. The overall current DE situation in the United States is summarized in Table 2-1 below.

Canada has about 15 DH systems in operation, with a total capacity of 910 MW$_t$. The largest systems are in Toronto (276 MW$_t$) and Montreal (100 MW$_t$). The systems are basically similar to those in the United States and are largely steam-based. In the last 10 years, small hot water systems were constructed in Charlottetown (Prince Edward Island), Cornwell and Windsor. The characteristics of the DE industry in Canada are quite similar to those in the United States, from the rise of institutional systems to the technical characteristics of the older and newer systems.

TABLE 2-1
District Energy Installations in the United States

	Urban & Community	College	Hospital	Industrial	Military	Other	Total
Number of DE Systems:	110	1,910	2,026	432	310	1,015	5,803
Value of Installed Plants (10^9\$):	6.7	51.5	74.8	11.7	27.1	5.5	177.3
Capacity (GW_t):	28	49	36	19	52	48	231
Annual Energy (PJ):	189	282	199	99	190	225	1,185
Distribution Line Length (km):	1,649	9,981	6,055	3,648	6,457	5,789	33,578

Urban Central Steam Systems. While there were about 250 urban steam systems in the United States after World War II, there were only about 60 systems in operation by 1980. The reason for the decline in DH systems was due to a number of factors. The practicality of higher and higher transmission voltages permitted inexpensive, efficient transport of electric power over many tens and then hundreds of miles, allowing power plants to be moved out of cities as they grew in capacity. This produced a shift to heat-only-boilers in the cities, and a decline in both fuel and economic efficiency. Also, as the electricity generating units remaining in the cities became less efficient as compared to the larger, more efficient plants now located outside of cities, the effective cost of the co-generated steam rose. This, in turn, caused the cost of thermal energy from urban steam DE systems to rise, leading to a steady loss of customers. These economic woes led to inadequate maintenance and failures to upgrade equipment, in turn producing worse performance and further erosion of the customer base.

These 60-odd urban systems still produce about 16% of the energy sent out by United States' DE systems, and the bulk of that energy is in the form of steam. From the early 1980's, rehabilitation of some old steam systems and development of new hot and chilled water systems took place. As a result of these developments, about 50 additional systems were constructed. Some of the newer systems send out hot and/or chilled water. A few of the most important are listed in Table 2-2.

Con Edison is by far the largest urban steam system in the United States. This system has been operating since 1882, with more than 100 miles of mains and service pipes serving more than 1,900 customers south of 96th Street in Manhattan, New York. It supplies space heating, domestic hot water and also the energy to run steam-powered air conditioning equipment for many of Manhattan's largest buildings, hospitals and other facilities. In April 1998, Con Edison proposed a long-range plan for restructuring the company's steam system that will encourage the development of new, modern steam-production plants to serve as a major source of the area's steam supply in future years.

A few other utilities, including Indianapolis Power and Light, Detroit Edison, and Wisconsin Power and Light also realize the value of their steam distribution systems for retention of customers against competing electricity and gas vendors and for demand management, especially by taking over cooling load. For the most part, however, the electric utilities backed out of the DE market by selling their steam systems to independent for-profit companies (such as Trigen and NRG Thermal). In either case, recent interest in fuel efficiency sparked by concerns about greenhouse gas emissions has led to renewed interest in DE systems, and there has been motion toward improved efficiency and system upkeep and improvement. The involvement of private ownership resulted in a substantial recovery of this industry.

TABLE 2-2
Large Urban Steam DH Systems in the United States

City	Company	Capacity	Annual Sendout	% Co-gen.	No. of Customers
		MW	PJ		
New York, NY	Consolidated Edison	3,516	30.1	53	1,960
Indianapolis, IN	I. Power & Light	440	6.57	5	280
Philadelphia, PA	Trigen P. Energy	615	4.20	70	390
Detroit, MI	D. Edison	410	3.39	38	310
Boston, MA	Trigen B. Energy	350	3.38	-	210
Minneapolis, MN	NRG Thermal	278	3.34	-	129
Milwaukee, WI	WI Electric Power	366	2.60	100	480
Lansing, MI	L. Board of Water & Light	309	1.22	-	300
Baltimore, MD	Trigen B. Energy	175	1.16	-	350

Institutional Systems. In the last few decades, a large number of institutions such as colleges, universities, hospitals and military bases have constructed and are operating their own DE systems. Such systems comprise the bulk of DE production in the United States. The energy sent out by these and the urban systems discussed above is shown in Table 2-3 below.

Institutional systems are also primarily steam systems, although there is an increasing trend toward distributing chilled water. Some of these systems generate their own electricity, which is included in the annual energy figures above. The military leads in chilled water production, reflecting the tendency of military bases to be located in warmer parts of the country. One company alone (The Pacific Energy Company of Commerce, California) produces 85% of the energy in the other category. Included in the institutional systems are the White House and other government buildings in Washington, D.C., which are served by a DH/DC system with a peak capacity of 6 PJ and a 12,000 ton chilled water plant operated by the General Services Administration. This system supplies 50 million square feet of office space to 112 government and quasi-government buildings.

Although these institutional systems have been built much more recently than the large urban systems, they were largely constructed using the same high temperature steam technology, with the same problems of high losses, complex maintenance and less than optimal economic performance. The percentage of hot water in Table 2-3 is an indicator of the level of penetration of more modern hot water systems in each sector.

TABLE 2-3
Energy Sent Out by American District Energy Systems by Sector

Sector	Annual Energy	Fraction of Thermal Capacity Going to:		
	PJ	Steam	Hot Water	Cold Water
Urban & Community	189	85%	10%	5%
College	282	71%	7%	22%
Hospital	199	69%	0.5%	31%
Industrial	99	62%	13%	25%
Military	190	31%	27%	42%
Other	225	12%	85%	3%
Total	1,185	75%	9%	16%

New Merchant DE Systems. The deregulation of gas and electric markets, discussed in the next chapter, has been accompanied by a flurry of activity in DE. All of the systems listed in Table 2-2 were originally owned by the local electric utility, which was either an investor-owned regulated entity or publicly-owned. As the economic outlook for the urban systems worsened, the utilities became interested in shedding what had become for them an onerous responsibility, and at the same time, private investors appeared who were able to purchase the DC or DH/DC system from the utility for a price that was often nominal and were also able to establish new operating companies facing less stringent regulation than that the utilities had operated under. The most active players in this arena are described in Annex 2.

Co-Generation and Fuels Utilized

As mentioned before, CHP or co-generation offers substantial benefits in fuel utilization and emissions and is the basis of many DE systems in the United States. CHP systems in the United States are supplied by a variety of combustible fuels, including gas, oil, coal, wood, other biomass, waste in resource recovery plants and even solar energy, although the latter is not currently cost-effective. Nuclear power is not used for CHP in the United States, because most reactors are by design far from population centers and also because the lower steam temperatures in nuclear plants would result in greater sacrifices in efficiency when heat is withdrawn above ambient temperatures.

DE systems are also supplied with thermal energy directly by boilers fueled by any of the fuel options mentioned above. In this case, efficiency and emissions improvements are derived from the more careful maintenance a large system receives and from the possibility of using fuels that would be inconvenient or unacceptable in individual heating systems, such as coal or biomass.

DE systems can also be powered by electricity or mechanical energy from a prime mover (motor). DE systems driven by electric resistance heating are seldom used. The type of annual fuel usage by all United States DH and DC systems is as follows:

TABLE 2-4

Fuels Used for DE Production in the United States in 1999

Natural Gas	41.1%
Coal	21.9%
Electricity (Mostly for district cooling)	20.8%
Oil	9.4%
Renewable	1.0%
Purchased Thermal	0.8%
Total	**100%**

Technical Aspects

Thermal Energy Production. The high temperature steam and hot water systems prevalent in the United States require production at up to 200°C. This can be achieved through co-generation, but at a substantial cost in electricity generation efficiency. Partly due to this, and partly to the construction of new generation plant far from urban areas, only a fraction of DE steam was produced by co-generation. For example, 15 years ago ten "typical" United States' systems, according to the weighted average of steam sent out, were producing 67% of their steam by co-generation. The rest was produced in HOBs at 70-80% fuel efficiency, producing a marked decline in economic efficiency. Since then, co-generation has increased significantly; the Edison Electric Institute estimates that total United States co-generation capacity increased from 10.5 GW in 1979 to 47 GW in 1997. Many of the new merchant systems are reviving or building co-generation systems based on gas-fired combustion turbines or combined cycle units.

Transport Medium. The classical medium for transporting thermal energy in America's older urban DH systems was high pressure saturated steam at 120–200°C (1–16 atmospheres gauge pressure). Lower steam pressures (0–1 atmosphere gauge) are used within buildings. Condensate return systems (made of much smaller pipe than the supply network) are preferable, since the treated water can more easily be maintained, but in many older systems were not installed, so condensate is simply dumped at the point of use.

Although steam carries roughly ten times the net heat of hot water (over reasonable temperature changes) per unit of mass, it is 100 to 800 times less dense and requires larger pipes even after higher flow velocities are taken into account. For this and other reasons, modern DH systems are more likely to use pressurized hot water to transport thermal energy. Low temperature hot water systems use supply temperatures below 120°C; designs between 120 and 175°C are considered medium temperature, and anything over 175°C is referred to as high temperature. Return temperatures are as low as is consistent with the service being provided, but rarely go below 70°C. All hot water systems are designed for return of the supply water.

District Heating Networks. The characteristic distribution network in the United States consists of insulated carbon steel pipe laid in concrete trenches. The pipe must be strong enough to withstand the pressure of the high temperature steam, making it costly. The networks are subject to a number of failure modes, especially when age, frost and vehicular traffic have cracked the concrete trenches so that water can infiltrate. At this point, water can infiltrate while the system is cool and produce external corrosion on the pipe, weakening it dramatically at the point of contact. The water is then vaporized when the steam returns. Leaving at high velocity, it can carry insulating material with it, leaving large gaps and heat leaks. Once these systems are more than a few years old, they experience losses, are expensive to maintain and are even dangerous (due to the possibility of a steam explosion).

Although preinsulated pipes, that can be buried directly in the ground and assembled with a minimum of on-site labor, are available especially from Europe, designers in the United States have been slow to make use of them. Lower temperatures (below 120-130°C) make possible the use of fiber-reinforced polymer (FRP) pipe, which is immune to chemical corrosion. This option, also, has not been widely adopted.

The older steam systems have heat losses ranging from 15 to 20% mostly because of deteriorated insulation. Most of the large old steam systems do not return condensate. As the systems are privatized, and as a result of the worst system sections having been shut down, the efficiency of steam systems is beginning to rise again. The hot water and DC systems are built to modern standards with factory insulated pipe, and losses are in the range of 5 to 6%.

Both hot water and cold water DE systems in the United States widely use variable flow control to match output to load and save pumping power.

Consumer Installations. Consumer installations in the United States can best be characterized as minimum capital cost designs. Steam is simply admitted through pressure reducers, and many hot water systems simply admit the transmission and distribution water into the building system. A few modern systems based on hot water use heat exchangers. Steam systems without condensate return never use heat exchangers; the steam is simply valved down to a lower pressure to match building requirements. Even with condensate return, they are generally used with steam supply only if hot water is demanded within the building. The result is a lack of control and poor ability to meet load accurately as well as difficulty isolating the system when repairs are needed.

Metering. Buildings served by DH systems in the United States are always metered. It is now widely accepted that actual metering of delivered energy is the best way to ensure efficiency within a DE system. If more than one economic entity is involved – for instance if a thermal transmission and distribution company is purchasing energy from an electric generation company – then metering satisfactory to both parties is necessary at each point where a transaction takes place. Also, the more metering is in place, the easier it will be to detect and locate insulation failures and leaks leading to losses. On the other hand, individual apartments are almost never metered, and sub-metering, where the building manager re-sells thermal energy to occupants, is not practiced.

Metering thermal energy involves measurement of both temperature change and fluid flow for water-based systems, and supply temperature or pressure and mass of condensate removed for steam. In both cases, several technical alternatives are available at varying costs and levels of accuracy. New developments in ultrasonic fluid velocity measurements promise to provide higher accuracy at lower cost if an adequate market develops.

Operation and Maintenance. Because the flow of steam is hard to modulate, control of operations is not precise and seldom efficient, since the only way to ensure customer satisfaction is to send out too much steam. Maintenance is a constant battle against corrosion and insulation failures in a decaying environment. These problems are present, although somewhat ameliorated, even for the newer steam systems. Only the small number of hot water systems based on prefabricated piping provide reliable and controllable systems.

A Transition to Modern District Energy Systems. A transition to hot water-based systems, and especially to systems designed around temperatures in the 100–130°C range, will increase the popularity of co-generation and the fuel efficiency of the DE systems in the United States. Shallow, accessible burial will lower maintenance costs of pipelines. Heat exchangers separating the transmission and distribution medium from building systems and often from the heat source are cost effective over system lifetimes and increase reliability and control even further. These technologies are available and well-tested from wide use in Europe. Unfortunately, designers have been slow to embrace the new technology, and developers have hesitated to rely on European suppliers of lower temperature transmission and distribution systems.

D. Central and Eastern European and the Former Soviet Union Systems

Extent of Coverage

The majority of the ten largest single schemes in the world are installed in the Russia and Central and Eastern Europe: St. Petersburg (66,000 GWh/year), Moscow (42,000 GWh/year), Kyiv (37,000 GWh/year), Warsaw (25,000 GWh/year), Prague (15,000 GWh/year), Minsk (12,500 GWh/year) and Bucharest (10,200 GWh/year).[5] The largest DH systems are included in Annex 3.

DH has a significant market share in the heating systems in Central and Eastern Europe and the FSU as portrayed by Figure 2-6 below.

---

[5] Koljonen, 1998; Ekono Energy, 1997.

FIGURE 2-6

Share of DH Households in Central and Eastern Europe and the Former Soviet Union in 1997

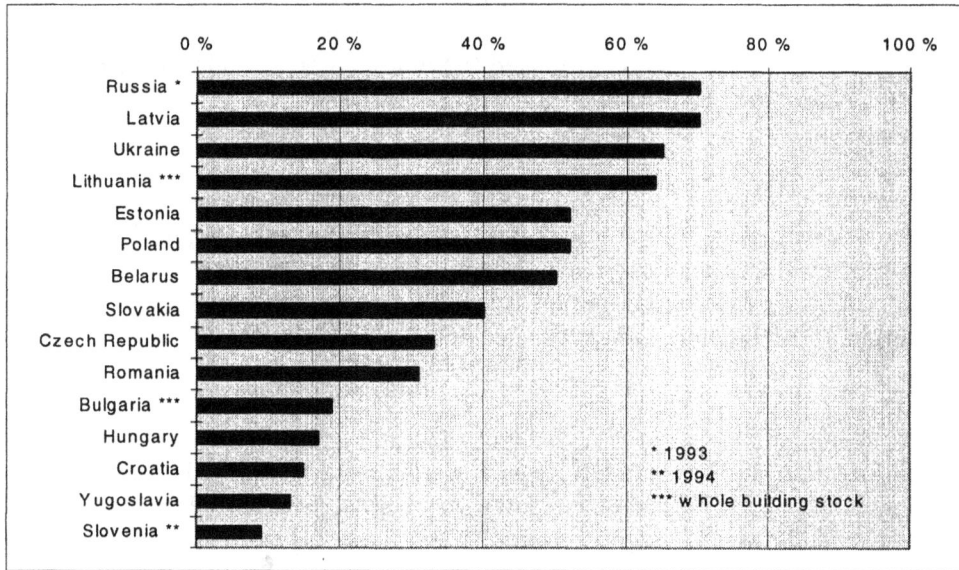

In Eastern and Central Europe and the FSU, most of the population resides in multifamily apartment buildings located in and around the cities. The domestic hot water load is an important component of DH. The domestic hot water load can regularly be about 20% of the total load and reach up to 40% at the peak use of domestic hot water.

The annual energy consumption and length of pipelines in some Eastern and Central European and FSU countries are shown in Figure 2-7 below. Data for Russia and Ukraine are not included in the figure because, due to their large schemes, differences between the other countries would be unreadable from the figure. From the information provided, it can be calculated that the energy carried per kilometer of network does not vary significantly among the Eastern European countries.

FIGURE 2-7

Annual Energy Consumption (GWh/year) and Length of Pipelines (km) in Some Eastern European countries in 1995-97

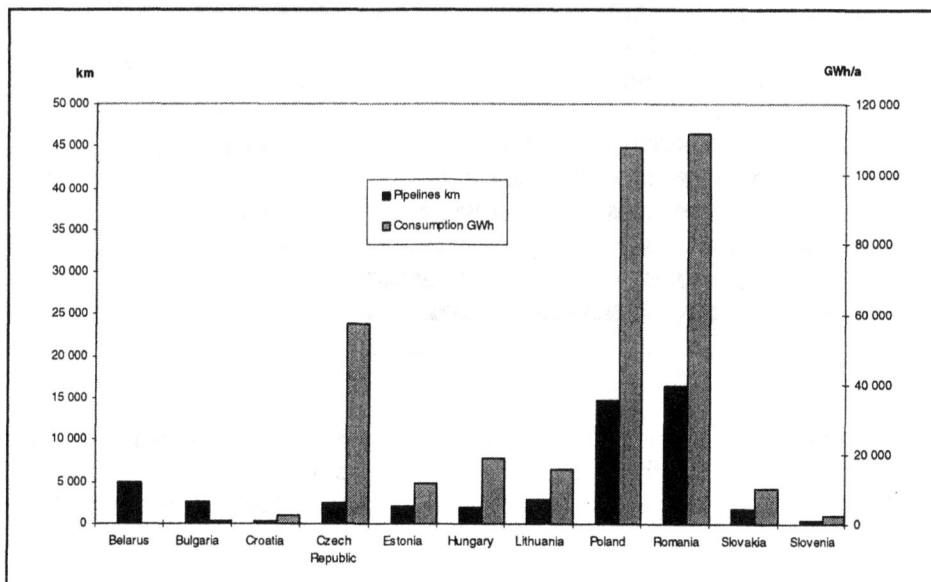

13

Co-generation

CHP is commonly used for DH production but is mainly available in DH systems in large cities. In most countries, however, reliable statistics are not available for the share of DH produced in CHP plants. In Russia, Hungary and Poland, the share is estimated to be around 50% while in the Czech Republic and Estonia, the share is estimated to be around 20-25%.

Fuels Utilized

Eastern European countries utilize a wide variety of fuels, as shown in Figure 2-8, but typically each country is highly dependent on one or two fuels. Natural gas is the most commonly used fuel and its share is increasing, but coal, lignite and heavy fuel oil are still common. The share of renewable fuels remains rather insignificant, as is the case for waste heat from incineration plants.

Technical Features

Technical Condition. Eastern European DH systems were mainly established during the influence of the former Soviet Union and are based on technology developed in the Soviet Union. After the transition process to market economies was initiated in the early 1990's, the DH systems in Russia and other Eastern European countries have been under a process of change. In general, Eastern European DH systems are in poor condition and in need of renovation and reconstruction. Huge investments are needed to lower operation and maintenance costs and to ensure that DH remains competitive with other heating alternatives.

FIGURE 2-8

Fuels Used for DH Production in Eastern Europe in 1997

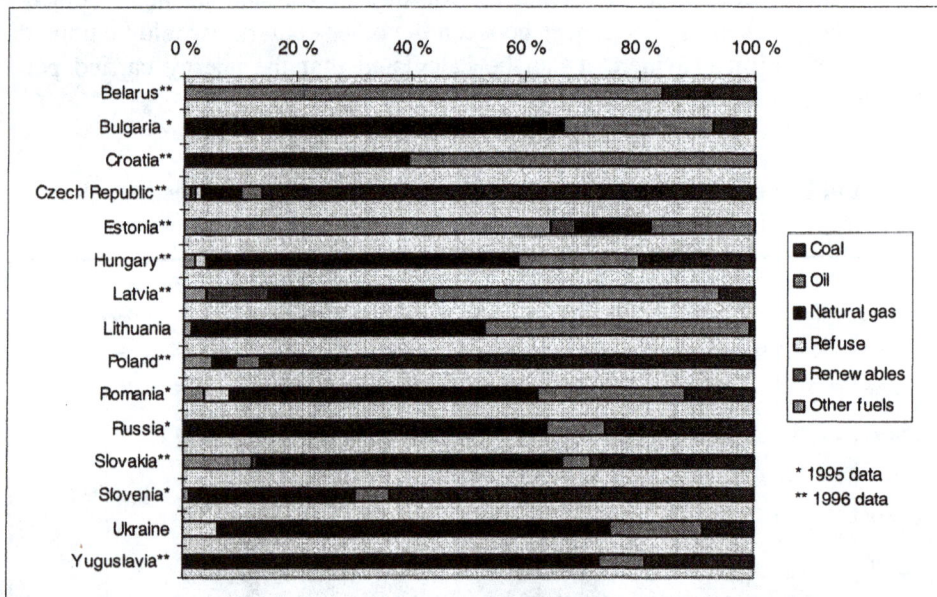

Heat Production. In the Russian and Eastern European schemes, DH is supplied from CHP plants and/or HOB plants. Where CHP is used, the peak and reserve capacity requirements are covered by HOBs. Typically, large DH systems have from one to three CHP plants and several hundred HOBs. As stated previously, a CHP plant is a technically more efficient way to produce heat and power than separated power generation and heat production. In Eastern Europe, the typical CHP plant efficiencies are around 70-75% as compared with 80-90% in Western Europe. The efficiency of the older HOBs in Eastern Europe are only 60-80%. However, with the introduction of modern automation and control systems, replacement of burners and cleaning of boiler surfaces, the efficiency can be typically increased to 85%. The efficiencies of new boilers are even higher, over 90%.

District Heating Networks. Typical factors leading to poor efficiency and various other problems in the DH networks in Eastern Europe include high levels of leakages, due to external and internal corrosion of pipes as well as insufficient pipe insulation, and the use of constant flow technology.

Network leakages are common due to both internal and external corrosion of pipes. It is not uncommon for water to infiltrate pipeline channels from the outside and high ground water to cause external corrosion when pipe insulation material becomes wet and ventilation in pipeline channels is poor. In systems with high water losses, make-up water has to be added. In the worst cases, the networks may have to be refilled a hundred times a year or more, as compared to 1-2 times per year in well-maintained Western DH systems. Where water treatment is not adequate, poor quality of make-up water corrodes the pipes from the inside. Heat losses are also typically high due to inadequate insulation. The thickness of insulation is less than in Western countries. There may also be problems with the quality of insulation which may not be homogenous but thick above and loose below the pipes.

The dominant mode of network operation in Eastern Europe has been constant flow regime. Basically, constant flow means that heat supply and heat demand are being adjusted by varying the flow temperature, typically in the range of 70-130°C, based on the ambient outdoor temperature. The adjustment of heat supply (and thus consumption level) in the typical constant flow DH system is carried out centrally at the heat production plants. Heat distribution to individual buildings depends entirely on the hydraulic balance of the network, leading to inaccurate heat distribution to buildings, e.g., too high indoor temperatures. In a constant flow system, each hydraulic section of the DH network system can be supplied typically by heat from only one location, which does not generally allow for the heat load to be dispatched from the least-cost production source.

Consumer Installations. Heat is transported through pipeline networks to substations which are needed to distribute the heat to consumers. Usually, heat is supplied both for space heating and domestic hot water. The substations may be located within the individual buildings or larger substations can serve a group of buildings through secondary networks, which typically involve 4 pipes – 2 for space heating and 2 for domestic hot water. The secondary networks usually experience high losses, and the technical lifetimes of these networks is short. The existing substations also experience heat losses due to insufficient installation. Both types of substations are used in the typical Eastern European schemes, while in Western Europe, most substations are installed in individual buildings.

Both direct and indirect consumer connections are used in Eastern Europe. As explained previously, indirect connection means that heat or domestic hot water is transferred by heat exchangers from the primary to the secondary network; the systems are thus hydraulically separated. Direct connection means that the water circulating in the DH network is introduced directly to the consumer installations. In systems with direct heat supply connections, the DH water flows into the building's secondary circuit (radiators). In systems with direct domestic hot water connections, the hot tap water is supplied directly from the DH pipes and needs to be made up at the point of supply. There are about 300 cities in the FSU which utilize direct domestic hot water systems. The advantages of indirect heat and hot water transfer include, for example, more efficient network regulation, better protection against corrosion and reduced need for make-up water.

Heat losses within buildings in Eastern Europe are typically estimated to be 25-40% higher than the design values.[6] The reasons for high losses are leaky windows and doors, uneven heat supply within buildings, and missing or insufficient insulation of the building basement and roof.

Metering. Virtually no heat or hot water metering existed before 1990 in residential, commercial and public sector buildings in many countries in Eastern Europe. There was very little point in installing meters because consumers could not regulate the heat supply. The lack of regulation and metering resulted in too low or too high room temperatures and further losses of heat from the opening of windows to cool the sometimes overheated rooms. Since 1990, many countries have made substantial progress in installing regulation and metering equipment. In many cities in the Czech Republic, Hungary and Poland, for example, the metering rate is close to 100%. By 1996, up to 50% of Lithuania's urban households lived in buildings with building-level meters. In Russia and Ukraine, however, very few (under 1%) residential buildings have meters.[7] In a few countries, such as Bulgaria, heat cost allocators are used in rehabilitated systems to allocate the building-level heating bill to the individual apartments.

Operation and Maintenance. Maintenance in Eastern Europe has typically concentrated on repairing damage that has occurred and not on preventing it in advance, although there are exceptions. Repair works of production plants and networks are usually carried out in summer, typically during a two to four week period. During this period, the water circulation in DH networks is totally shut off, with the result that consumers do not obtain even domestic hot water. In most Central and Eastern European and FSU systems, DH networks are tested by pressure once a year to reveal weak pipelines and leaks. For example, in Kyivenergo's systems (Kyiv, Ukraine), this has proven to be efficient because typically about ten breakages are repaired during the heating season while some four hundred are repaired outside the heating season.

Monitoring and control of the CHP plants, HOBs and pumping stations in Eastern Europe is often carried out from a central headquarters by telephone. In most cases, the supplied energy is measured only at the CHP and large HOB plants. Energy management systems do not exist. Operation is monitored afterwards by checking manual reports of heat supply from heat production units, and manual operation requires substantial manpower. Today, however, an increasing number of on-line monitoring systems are being installed.

The key characteristics of DH in the above Central and Eastern European and FSU countries, as well as in Armenia, Georgia, Kazakhstan and Kyrgyz Republic, are described in more detail in Annex 4.

E. Asian District Energy Systems

Extent and Key Characteristics of DH in Asian Countries

Peoples Republic of China. By the end of 1997, 278 of the 668 largest cities in China had DH facilities. The DH coverage is about 12% for the whole country and 29% for the large cities in the northern part of the country. Both hot water and steam are used as heat transfer media. Over 25,000 kilometers of the pipelines transfer hot water and about 7,000 kilometers transfer steam. The installed heating capacity is about 70,000 MW of hot water and 65,200 ton/hour of steam. Only units with a capacity over 7 MW are regarded as DH sources in the national statistics.

Coal is the main energy source in China; in 1995 the consumption was about one billion tons of coal equivalents. Coal is used in the heating sector both for DH and in individual boilers in houses. Even in

---

[6] UNDP and ESMAP, 1998.
[7] UNDP and ESMAP, 1998.

urban households, coal is commonly used for cooking and space heating. In rural households, 90% of the energy consumption is biomass. In urban areas, use of electricity and LPG are being used increasingly. Natural gas is available only from a few locations, and oil is preferable.

Japan. The history of DE systems is very recent in Japan. The first network was installed at the Universal Exhibition at Osaka in the late 1970's. During the last ten years, the systems have increased rapidly, though the cost of distribution is very high. DC represents over 40% of the energy furnished through the 120 DH/DC schemes installed.

Republic of South Korea. In the early 1980's, the Government began to implement a widespread CHP and DH scheme in Seoul and its suburbs, to respond to the increased energy demand in the area due to rapid urbanization and severe air quality problems. Most of the schemes are concentrated in Seoul and its suburbs. Currently, the share of CHP in DH production is high because all the schemes have been developed around a CHP plant, and the interconnection of the schemes is being carried out. In 1997, DH had an 8% market share in the residential areas. The anticipated growth is to reach a 15% market share by 2001.

Mongolia. Mongolia has a population of about two million of whom about 615,000 live in the capital city, Ulaanbaatar. Central Energy System (CES) is the main energy supply utility, providing DH and industrial steam in Ulaanbaatar, Erdenet and Darkhan. The DH supply was about 6 TWh in 1995. The number of residential consumers in Ulaanbaatar was 45,000 flats (housing 260,000 people) and 1,800 other buildings. Heat is produced mainly in three major coal-fired CHP plants, and, in addition, some coal-fired HOBs are used. In urban areas where houses are not connected to the DH network, coal is usually used for heating, while in other areas, mainly wood is used.

3. Institutional and Regulatory Arrangements for District Heating

A. Western Europe

Institutional Aspects

Energy Policy. The successful development of DH in Western Europe was driven, at least originally, by reaction to the oil shocks of the early 1970's and the realization that energy security could only be ensured by reducing dependence on imported products. Moving into the 1990's and the present millennium, an equal driver has been the recognition of environmental crises in the making and of the major role that DE can play in averting or minimizing them.

Most of the Western European countries belong to the European Union (EU) which has taken a very positive attitude towards the development of CHP and DH. The EU has promoted the concept of CHP since 1974. In 1977, the Council recommended that member states establish advisory bodies and committees to encourage CHP and heat transport schemes.[8] In 1988, another recommendation encouraged the removal of legal and administrative obstacles in the co-operation between public utilities and electricity auto-producers which would use renewables, waste fuels and CHP.[9] In 1992, obstacles for CHP development were still found. The main problems identified were the still undeveloped relationship between auto-producers and electricity production utilities, although the situation had somewhat improved, and the lack of progress in achieving a competitive internal market in electricity in EU member states.[10]

Since 1992, additional steps have been taken which support the development of DH and CHP use. The directives concerning the liberalization of the internal electricity and gas markets became effective in 1996 and 1998, respectively. A new proposal for restructuring the community framework for the taxation of energy products was adopted in 1997. This framework offers member states a possibility to grant fiscal advantages to renewable energy sources and co-generated heat.[11]

DH and heat production are subject to three different kinds of taxes: (1) value-added tax (VAT); (2) special fuel excise taxes; and (3) environmental taxes. VAT may be flat (e.g. Denmark, Finland, the Netherlands, Norway and Sweden) or may be lower for DH (e.g. Italy). If fuel excise taxes are applied (e.g. Denmark, Finland, France and Sweden), they typically are lower for biomass and natural gas to encourage greater use of these preferable fuels. Environmental taxes, usually based on carbon and/or sulphur content of fuels, are applied in several countries.[12] The environmental benefits that can be achieved through the use of renewables and CHP can be supported by using such taxes.

The 1995 White Paper "An Energy Policy for the European Union" outlines three central tenets: (1) competitiveness of European businesses in the global markets; (2) environmental protection; and (3) security of supply. CHP is considered an important and cost effective contributor to all three. In the White Paper, the Commission committed itself to present a strategy offering a coherent approach for the promotion of CHP within the EU. The strategy was presented by the Commission to the Council and the European Parliament in 1997. A key finding of the strategy is that CHP is one of the few technologies that can offer a significant short or medium-term contribution to improving energy efficiency in the EU. CHP is also recognized as making a positive contribution to the environmental policies, especially in fulfilling

---

[8] 77/714/EEC.
[9] 86/611/EEC.
[10] European Commission, 1992 and 1997.
[11] COM (97) 30 final; European Commission, 1997.
[12] Euroheat & Power, 1998.

the Kyoto requirements,[13] as discussed later. The strategy aims at doubling the CHP market share from 9% to 18% of gross generation by 2010 in the EU. Ambitious plans have been presented (900-1000 TWh of electricity from CHP, reduction of carbon emissions by 300 million metric tons per year which is 9% of projected EU emissions in 2010).

The EU strategy identifies several economic, regulatory and institutional barriers to increasing the use of CHP. The strategy concludes that many of the important barriers still result from the relationship between co-generators and electricity production utilities. Obstacles to access to the grid, inadequate payments for sales of surplus capacity to the grid, and high tariffs for stand-by and top-up supplies are key factors even in the partly liberalized European energy market.[14] Experience in the United States and Great Britain over the next few years should offer some sense of how true this is, since these barriers have been or are in the process of being dramatically reduced as deregulation is implemented in these countries. The strategy also identifies specific barriers for DH. One such barrier is related to the trends in energy consumption, which indicate an increasing use of electricity but a stagnant market for heating, partly due to improved building standards and insulation.[15]

The EU gives recommendations and has some authority over the member countries, but each member country formulates its own national energy policies. Typical goals include energy savings, improved energy efficiency both in production and use in all sectors, use of indigenous energy sources, increased use of renewable energy, diversity in use of fuels, security of supply and addressing emergency issues and environmental acceptability. The priorities vary country by country, but all EU countries stress the importance of energy savings and improved energy efficiency. DH and CHP have been considered as important tools in the realization of energy policies. In Denmark, Finland and Sweden, the Government and local authorities have taken a positive attitude towards DH. DH, together with CHP, is considered to have made a significant contribution to energy savings efforts. Germany and Italy share the same attitude.

Other policies promoting DH exist in the individual member states. For example, in Italy, electricity production from a CHP plant can be sold to the National Electricity Board at prices that encourage auto-producing. In Germany, an energy savings program was created in 1977 to promote energy savings in buildings by applying lower assessed values of properties for tax purposes for new technologies including DH. In Finland, DH has not received any direct financial or legislative support, but has been viable on its own merits, including the use of renewable and indigenous fuels having lower environmental and fuel taxes, such as wood, which are aimed at energy saving or advancing the use of domestic fuels. In Sweden, the Energy Agreement of 1997 entailed the earmarking of funds to subsidize new electricity generating capacity, and for CHP based on bio-fuels, about US$ 50 million was earmarked.

Ownership of Heat Production and Distribution. In Western Europe, DH, electricity and gas businesses traditionally have been integrated in the same company or are organized under the same ownership, such as state, regional or municipal. For example, in Germany, the supply of DH in some locations is carried out by regional electricity supply companies, companies in municipality possession that may have a regional focus, gas industry, mineral oil industry and coal industry. In Germany, the oil and coal industries manage more than half of the present existing DH networks. The supposed benefits of this organization have been the possibility for overall optimization of energy supply, limited competition of energy forms, and coordinated promotion of energy saving information to consumers in order to reduce specific consumption.

---

[13] European Commission, 1997.
[14] European Commission, 1997.
[15] European Commission, 1997.

19

In Northern Europe, traditionally DH has been produced and distributed by municipal utilities. The DH utilities operate according to business principles either as a city department or as limited liability companies. The DH business is characterized by a large number of small companies within each country. The enterprises have not been forced to merge, form large regional units nor have been nationalized and restructured into one national public utility, as the case has been sometimes with the electricity supply industry. Another typical characteristic is that DH companies are both producers and distributors of heat, so that they can optimize their operation in order to compete with decentralized heating systems. This is the case especially in Scandinavia. In some cases, bulk heat is acquired from nearby industries or power plants.

Today, there is a clear tendency in Western Europe towards unbundling of activities in companies with both DH and electricity and other business activities and more private ownership, although public ownership is still the predominate ownership form today. The aim of separating DH and electricity businesses is to improve heating market competitiveness and efficiency, to prevent cross-subsidies from one business activity to another, and to improve transparency of costs and profitability. DH businesses typically include production, transmission, distribution and supply within one company in order to be able to optimize the DH system so that it can better compete with decentralized heating alternatives.

Regulation

Most of the Western European DH utilities are considered to be "natural monopolies" in their service areas, having reached a dominant market position as a consequence of their development. However, DH is not a monopoly in the true sense, since DH is operating in an inherently competitive environment whereby decentralized heating options, such as individual boilers based on various fuels, are always available. Furthermore, the laws do not include any stipulations that DH would have any monopolistic position. DH in Scandinavia, as well as in some other parts of Europe, is not subject to regulation by regulatory authorities, as it is judged to be regulated by the market. However, the dominant feature of DH has led to the close scrutiny by consumer protection authorities to oversee DH in order to prevent the misuse of DH in its dominant position in the given market.

In some Western European countries where regulatory authorities have jurisdiction over DH pricing, regulation aims at preventing misuse of the stated market position. This typically includes the following: (a) price level is "reasonable" (as defined by the market); (b) pricing "corresponds" to the costs; (c) same kind of customers are treated equally; (d) in pricing different energy products, such as heat and electricity, from the same company, the pricing policies of the individual products shall not be inter-linked (such as when discounts are given when several products are purchased from the same company); and (e) cross-subsidies are prohibited between different business activities of the same company. Usually more than 50% of the DH tariff is due to the fuel costs. When the fuel prices either increase or fall, the DH enterprise usually may adjust the DH tariff without any separate approval from the regulator.

The Impact of Deregulation of Electricity Markets

Deregulation of electricity markets is having an impact on DH and co-generation. Own production capacity has become strategically important for the electric utilities, and DH with CHP provides access to low-cost electricity generation which helps to improve the electric utility's market position. This has increased the interest towards CHP schemes in some countries, such as the United Kingdom. On the other hand, in some European countries such as Germany and Denmark, energy sector restructuring may result in stranded CHP and DH assets. In the past, there was a tendency to overextend DH systems due to the practice of allocating benefits from CHP in the opposite way as in Eastern Europe, i.e., by allocating all the benefits to heat. This makes heat artificially cheap and, conversely, electricity from CHP artificially expensive. Those CHP plants may lose many of their electricity customers unless they lower their prices, but this would require raising tariffs for heat customers.

In addition, the risk of undertaking new investments has increased, because the previous supply monopoly areas do not exist anymore. Poor investment decisions cannot be financed by raising electricity sales prices, because the supply is subject to competition. At the same time, the future development of electricity prices is unclear. In the case of new capacity construction, the prices should be high enough to cover the long-term marginal costs of the new capacity. The market is some times, however, distorted by existing contracts which are based on short-term marginal costs, affecting spot prices to be lower than long-run marginal costs. Therefore, the construction of new capacity may be delayed from the optimum.

International liberalization of electricity trade as well as increased cross-border transmission capacity from Scandinavia to continental Europe should decrease the electricity price differences between European countries. In many European countries, some co-generators fear that they will not survive on the open electricity market. This will be the case if the scheme itself is uneconomical and has been based on a monopoly supply position or subsidies. On the other hand, in some European countries, this fear does not generally exist since the schemes have traditionally been based on competitive principles. Investments have only been made in such cases where the economic viability compared to other options has been proven.

Therefore, the impact of electricity market liberalization on DH and CHP is greatest in countries where the markets have been most monopolistic and especially where the conditions for auto-producers have been non-existing or very unfavorable. This has prevented the development of medium and small-size local DH/CHP schemes. In Scandinavian countries, on the other hand, CHP was well developed already during the "unliberalized" time, and deregulation brought little or no change in this respect.

One of the key points concerning the deregulation of electricity markets is the failure of those markets to reflect the cost of environmental externalities (from acid rain to carbon emissions). The Commission proposes various possible limitations on electricity markets to ensure that these externalities are in fact accounted for. These measures would include tax incentives and portfolio requirements for purchase of a minimum fraction of annual energy from CHP plants burning biomass. Although these measures might appear to be subsidies to some observers, the Commission holds that as compensation for very real externalities, they would be acceptable under current trade treaties.

B. United States
Institutional Aspects

The technology of DE in the United States has been shaped and changed by the institutional structures within which it developed. From the 1930's, DH in the United States became subject to state regulation as a public utility. Following a period of "benign neglect" while the urban systems fell into disrepair, the energy crisis of the 1970's provoked a flurry of interest from the US Department of Energy as it promoted DE and CHP as a means to greater fuel utilization efficiency. At the same time, the idea of deregulation of the utility industry was taking hold, and the pro-market policies of the 1980's precluded any major government investment in DE systems. Finally, the arrival of deregulation has had a profound and positive impact on the DE industry. These events are examined below.

Ownership of Heat Production and Distribution. Electric utilities first began selling thermal energy in urban areas as an incentive to attract potential customers who were already supplying themselves with individual, building-level CHP, as discussed previously. In doing so, they made use of the excess heat produced by their own generation stations. To provide a full service solution to their customers, they also obtained the rights-of-way and laid transmission and distribution pipelines. Consequently, from its inception, urban DE was a system where one entity, usually the electric utility, owned both the production and the distribution facilities for electricity and heat. Even companies dedicated solely to DH owned both production and distribution facilities, since it was the easiest way to ensure compatibility and coordination of supply and distribution.

Interactions of Institutional Systems with Utilities. Institutional DH systems, such as hospitals, universities and military bases, originated as expansions and improvements to original heating systems based on boilers in individual buildings. Initially they were simply central boiler-fired steam systems, taking advantage of the increased efficiency and control made possible by using one large boiler. As such, they had very little interaction with utilities, other than to purchase electricity and, where available, gas.

However, as institutional systems became larger and more widespread in the 1950's and 1960's, the advantages of co-generation became apparent to utility plant managers. Foreseeing a possible loss of considerable load to the institutional systems, the utilities often forced institutions that developed co-generation to run on a stand-alone basis. When a residential project in Queens, New York turned to co-generation, Con Edison not only refused to provide back-up power, but dug up and removed several hundred feet of cable to emphasize that the project was on its own. During the 1976 blackout, its lights were conspicuously burning.

In trying to prevent institutional systems from developing co-generation, the utilities were fighting a losing battle. The military and large industrial plants had the power to proceed on their own, and by 1978 the law began to catch up with technology.

Regulation

As with the electric and telephone industries, DH came to be considered a "natural monopoly" in the early part of the last century and was brought under the same sort of regulatory control as the other utilities. The reasoning behind this was straightforward: it made no technical sense to have two or more intertwined systems of electric distribution lines, telephone wires or steam pipes. Such a melange would be duplicative and would cost society much more than one united system. Hence the "natural monopoly" concept.

However, once the monopoly nature of these types of utilities was acknowledged, concerns were raised as to how to prevent rampant abuse and price gouging. Two answers emerged, which have continued to co-exist in the United States up to the present day. The first was public ownership of the utilities, usually either on a municipal level for electric or steam distribution and on a federal level for large generation projects, such as hydropower. The second was to allow the utility to function as a private corporation, usually owned by stockholders, but be subject to strict regulation, normally emanating from the state government.

DH systems emerged under both models. In large Eastern cities like New York, Boston and Philadelphia, the steam system was a component of the local, privately-owned, publicly regulated electric utility. In broad terms, and in the same way as the electric and gas divisions of the utility, they were allowed to raise rates to make a "reasonable" profit for their shareholders. That profit was calculated as a percentage of the capital invested in the steam system, known as the "rate base." The utility had to justify to the regulators that investments in the steam system, additions to the rate base, were "used and useful," by showing that additions were necessary to serve legitimate customers and that improvements would result in lower maintenance and operating costs. This criterion was designed to prevent price gouging and unjustified expansion of the system.

This system of regulation worked reasonably well when the steam (and electric) utilities were expanding their customer bases up through the 1940's and early 1950's. However, regulation was never enforced as rigorously for the steam systems as it was for the electricity side, since the monopoly concept as applied to DE was not as clear, given that competition from other heating options existed. For electricity, the states passed laws actually granting the utilities a monopoly position and made it illegal for anyone else to sell electricity or obtain rights-of-way for that purpose, so the electric utility really had no competitors. A few co-generation projects were created, but because the utility would not provide them with back-up power they had to achieve very high reliability on their own, and this was expensive.

The DE steam system, in contrast, had a very real competitor, in that any new building could choose to install boilers burning either gas or fuel oil. Buildings which had already been designed and constructed around the provision of municipal steam, however, had much more limited options. They had been built without boiler rooms or smokestacks, and the cost of adding these facilities later was much higher than it would have been during construction. The result was a collection of older, downtown buildings in each city that were trapped in the aging steam system: steam prices could rise well above the costs of individual building-level boilers based on oil or gas, and it was still not cost-effective for the owners of older building to back out. With many costs fixed and new buildings avoiding the expensive steam system, the price of steam rose steadily, leading to further erosion of the customer base. The laxity in regulatory oversight contributed to poor decisions such as the abandonment of co-generation in favor of heat-only-boilers.

It would not be fair to blame the regulatory system for the decline of large urban steam systems in the United States. Many technical changes had negative impacts making steam less efficient for DE purposes. It probably is fair to say, however, that the regulatory system was not responsive to the changing circumstances and made flexible reactions by the utilities difficult. Being exempt from the regulatory system did not guarantee success either; many municipal DE systems suffered comparable decline, while others did not. The flexibility possible in the case of municipal ownership was only helpful if a motivated and resourceful government was available to take advantage of it.

The Impact of Deregulation on Traditional Urban District Energy Systems. The history and status of the deregulation of the gas and electricity markets in the United States is presented in Annex 5. The deregulation of the electric utilities has had various effects on DE systems in different contexts. Some electric utilities that had been operating urban steam systems have elected to continue to do so. Con Edison in New York and Indianapolis Power and Light are the two largest; others include Detroit Edison, Wisconsin Power and Light in Milwaukee and the Lansing (Michigan) Board of Power and Light. These systems, constituting at least one-third of United States' urban steam capacity, will remain under the regulation of their public service commissions (PSCs), including approval of their rates. As part of the deregulatory activity, the PSCs are showing more flexibility, and, for example, are permitting the utilities to sign on new customers without regulatory oversight of rates.

Deregulation will also pose new challenges, however. As electric utilities sell generation capacity, heat contracts must be maintained with new owners and the need for thermal energy must be coordinated with the independent system operator's (ISO) dispatch priorities. In New York City, the sale of one plant required a long-term contract for the delivery of thermal energy to Con Ed, including restrictions that the plant be operated to meet thermal demand, which may lead to less-than-optimal electricity generation.

Allocation of generation expenses between thermal and electrical energy has always been problematic where regulators request a cost-based allocation, which is not possible in joint production processes such as CHP. In cases where the joint products are subject to regulation, this allocation has to be made according to some principles. Some charges, such as those associated with extra steam extractors or heat exchangers required for DH or with the generators themselves are easy to allocate. Allocation of other charges, such as fuel, is done according to many time-honored but arbitrary procedures approved by the regulators.

The Impact of Deregulation on Institutional Systems. The impact of deregulation on institutional systems is almost entirely beneficial. It places the institutional co-generators firmly in the wholesale market for regular supplementary power or emergency back-up, allowing at least the larger co-generators to shop with considerable bargaining power. Deregulation also puts them in the wholesale generation market with their surplus power, which may be a substantial asset in urban areas such as Manhattan with restricted transmission access and limited generation capacity. Under procedures now being implemented in New York State, energy and capacity are being sold independently. Of course, utilities can only sell available capacity, and if they need back-up energy, they must pay for the associated installed capacity (ICAP).

Deregulation and Merchant District Energy Systems. As discussed previously, many independent companies are buying up urban steam systems and reorganizing and reinvigorating them. One step making this possible has been the decrepit, sometimes nearly defunct, state of the existing systems. Given the threat of collapse of the existing DH system, merchant companies such as Trigen have been able to negotiate with the PSCs as well as the parent companies to lessen or even remove the regulatory burden. In most cases they have won relief from price regulation, arguing that the availability of competitive, individual building systems would preclude monopolistic pricing. In all cases, they have won relief from price regulation for new customers. The result has been the encouraging burst of activity.

Of course the absence of regulation carries with it the absence of security. When regulated, the utilities were guaranteed a return on their "rate base." A merchant system is guaranteed nothing except what it established through contracts with customers. How the risk of, for example, unexpectedly rising oil prices, is allocated between the merchant DE system and its customers is a matter for contract and tariff structure. However, to present a package that will be more attractive than individual boilers, the merchant DE system must assume a substantial part of this risk. For this and many other reasons, deregulation presents the merchant DE systems substantial dangers as well as offering them significant rewards.

C. Eastern Europe

Institutional Aspects

Energy Policy. In Eastern Europe, DH has been a preferred, in many cases even mandatory, heating form required by the central planning authorities. For example, the 5-year plan of 1920 for country-wide electrification in the FSU introduced the idea of CHP and DH; the 5-year plan of 1933-37 required a substantial increase of DH systems. In Romania, the 10-year plan of 1950 introduced an expansion of CHP and DH. The standard design of cities in Eastern and Central Europe and the FSU included centralized heat and hot water supply systems. The high-rise apartment buildings where most of the families live are supplied by heat and hot water from DH systems.[16]

The energy policies applied today throughout Eastern and Central Europe and the FSU reflect the need for introduction of energy efficiency and conservation measures and a reduction in energy intensity, especially in end-use applications. In addition, the energy policies stress the importance of energy market liberalization, security of supply, diversification of fuel use, increased use of indigenous and renewable fuels, and economic efficiency combined with social acceptability. Some of the Eastern and Central European countries plan to join the EU in the near future, and therefore are in process of harmonizing their energy sector policies with that of the EU. Some of the largest sources of energy wastage in Eastern and

---

[16] UNDP and ESMAP, 1998.

Central Europe and the FSU are the outdated and badly maintained DH systems, including CHP plants. Reducing losses in DH systems and increasing the efficiency of CHP generation are being promoted in the national energy policies of most countries in this region.

At this time, tax policies encouraging energy savings and environmental improvements in Eastern and Central Europe and the FSU are in the early stages of development. DH and heat production are usually subject only to value-added tax (VAT). VAT may be flat (e.g. Czech Republic, Poland and Romania) or favorable for DH (e.g. Estonia and Slovenia). Fuel taxes and environmental taxes are usually not collected. Slovenia has been the first country to apply an environmental tax on CO_2 emissions from 1997. Some countries, such as Poland, apply emission charges. Generally, environmental taxes in Eastern Europe, where levied, are low.

Ownership of Heat Production and Distribution. The Eastern and Central European and FSU DH utilities can be owned by the state, regional authorities or municipalities. In this region, DH supply is traditionally based on a two-company structure in the large cities. CHP and large HOB activity and heat transmission networks are normally operated by one company, known as "energo" in Russia and the FSU states, and the operation of distribution and isolated networks and small boilers by another company, usually municipally-owned. In addition, a number of small separate networks are operated by institutional or industrial entities. The ownership and control of energos varies. Typically, 49% of each energo in Russia is owned by the federal power company, RAO EES Rossija, with the balance split among the employees and private shareholders. In other countries of Eastern and Central Europe and the FSU, the energos are typically state-owned.

In this region in a number of cities such as Tallinn (Estonia), Riga (Latvia) and Kiev (Ukraine), the two-company structure is being consolidated into one company serving the DH consumers within a city. This allows for optimization of the DH system so that the resulting company can better compete with the emerging decentralized heating options. In a number of cases, the CHP plants are still owned and operated as separate entities and this leads to difficulties for the long-term optimization of the DH system as a whole.

The technical and financial performance of the DH systems in Eastern and Central Europe and the FSU is, in most cases, too poor to attract private sector investments. In some cases, however, private investments have recently started. As an example, a Finnish power company has purchased 60% of the CHP company in Budapest, Hungary, where the heat distribution systems are still owned by the municipality but privatization is planned for the near future. A UK power operator has invested in heat and power systems in the Czech Republic, and the state electricity monopoly of France has invested in the CHP/DH system in Krakow, Poland. One of the first independent power projects in the region is in the Czech Republic, the Energy Center Kladno, which sells electricity to the state power company and heat to the local DH network.[17] Further examples include an American-Israeli joint-venture which completed the acquisition and takeover of four CHP stations in Eastern Kazakhstan, an American-Dutch joint venture which acquired a majority interest in a wholesale power and DH company in the Czech Republic, and an American-Czech joint venture which acquired a majority interest in two Czech energy centers.

Regulation

Regulation of DH is Eastern and Central Europe and the FSU is typically under the jurisdiction of the municipalities, and regulation of electricity from CHP plants is typically under the jurisdiction of federal or state authorities. Some countries are establishing independent regulators mainly for electricity activities. DH has traditionally been viewed as a "natural monopoly," as it has been in many other parts of

---

[17] Power Economics, 1998.

the world, and has been brought under the same kind of regulatory control as other utilities. Regulation usually attempts to establish prices for heat on a "cost plus" basis, and the concept of pricing of heat based on its market value is still a novel idea.

The allocation of expenses in the joint production of heat and electricity in CHP plants in this region has generally been carried out according to administrative principles which allocate the benefits of the joint production to electricity rather than attempting to share the benefits with the two products. This has resulted in prices for heat from the CHP plants to be at the same level as heat produced in HOBs or even higher. The deficiencies of this pricing method are being recognized in a number of countries in this region, especially as consumers now have other alternatives to DH. Changes to the allocation of expenses in CHP plants are being introduced in a number of places, such as Poland and Kiev, Ukraine.

4. Economic and Financial Aspects of District Heating

A. Economic Benefits of District Heating

Lower Cost

The competitive advantage of DH is its ability to produce heat at a lower cost and in a more environmentally friendly way than with individual building boilers in areas of high heat load density. Thus, the extent of heat produced in co-generation and the heat load density are crucial factors to consider when making investment decisions in DH. Several case studies confirm the general superiority of DH systems in densely populated areas supplied by co-generation facilities as compared to other decentralized space heating options.[18] A high level of fuel savings must be obtained or otherwise building boilers will have an advantage.

The disadvantage of DH as compared to individual building boilers is related to the cost of transporting heat from the centralized heat production plant to the consumers. If the comparison is made with individual gas-fired building boilers, the cost of transporting heat would be higher than the cost of transporting gas. Therefore, in order for DH to be competitive, the level of fuel savings would have to more than offset the high costs of transportation to the individual consumers. The density of heat demand per kilometer of DH pipeline is an important indicator for the cost of transportation.

The World Bank has developed a simplified model for determining the least-cost heating options based on Western technology in a green field situation of centralized heating as compared with decentralized heating. The model compares three green field options where no sunk costs for any equipment or pipeline exist: (a) centralized option where heat is produced in gas-fired HOBs, (b) centralized option where heat is produced at zero cost,[19] and (c) decentralized option of individual gas boilers where it is assumed that existing gas networks reach every building. The model compares the long-run costs at the user level of these three options.

This model was applied to the case of Kiev, Ukraine using 1997/98 data,[20] whereby the proposed rehabilitation and expansion of DH heat production capacities at various boiler plants was compared to the option of decentralizing heat supply to buildings by installing individual gas boilers, and the results are shown in Figure 4-1 below:

---

[18] ESMAP, 1998.

[19] Used for comparison purposes for providing the absolute lower limit below which no centralized heat source (whether based on inexpensive or waste fuel, co-generation, free heat from waste incineration or CHP, etc.) can be less expensive than heat from individual boilers in buildings.

[20] The model assumes a capital cost of 10%, technical life of 20 years for all options, centralized boiler cost of $ 60,000/Gcal/h, DH network and substation cost of $ 487,000 per km per Gcal/h, gas network cost of $12,000/Gcal/h (including reinforcing of the present gas network for building-level boilers, typical (300 m) connection lines to buildings, pressure regulators and gas meters), individual boiler cost of $116,000/Gcal/h (including installation and construction work required for two boiler units to achieve the required reliability and to be consistent with the building code), DH network losses of 8%, operating and maintenance costs of 1.5% and 3% of capital costs for centralized and decentralized heating systems respectively, a gas price of about $ 82.4/1,000 m3 or $10.20/Gcal, and a peak load duration of 2,200 hours.

FIGURE 4.1
Centralized Versus Decentralized Heating Options

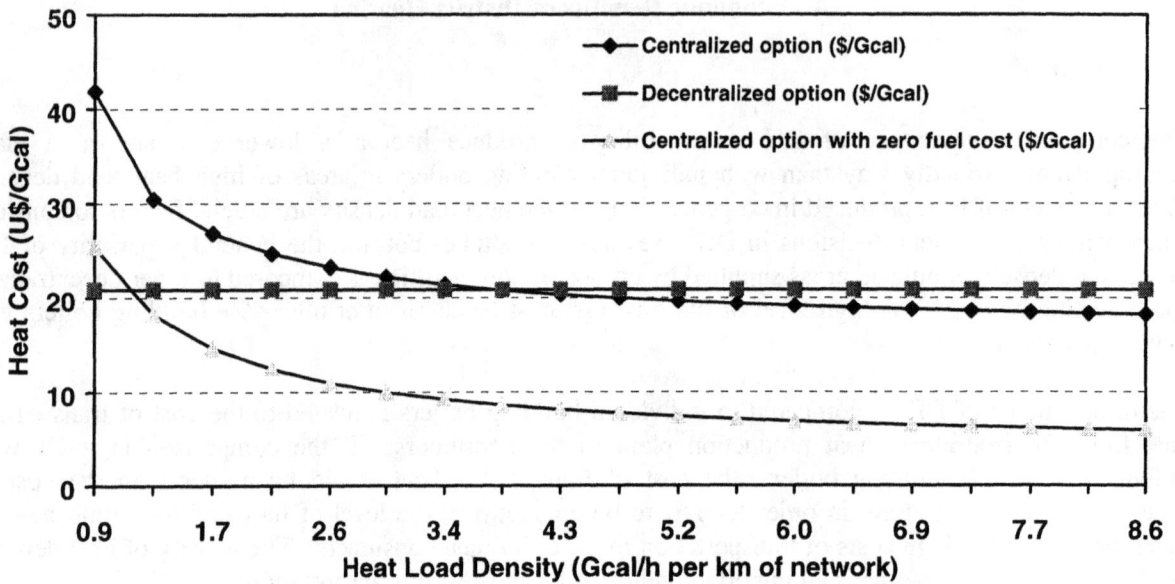

The graph above shows that, for high head load densities of more than 4 Gcal/h per km of network, centralized DH production at HOB and/or CHP plants is the preferred option. Where heat load densities are between 1 Gcal/h per km and 4 Gcal/h per km, a detailed analysis would be needed using actual fuel and heat production costs at the plant to determine the preferred option. Where heat load densities are below 1 Gcal/h per km, decentralized individual boilers are the preferred option. Since the average heat load density in the main interconnected DH system in Kiev is 5.9 Gcal/h per km of network, DH is the least-cost option.

However, there are other situations where the least-cost option is not DH. In Orenberg, Russia, for example, the least-cost heating option is building boilers. The determining factors are the low cost of natural gas and low long-run marginal cost of electricity. Orenberg is located on top of low-cost, natural gas resources, which are too far from the borders of gas-importing countries to be exported at a reasonable cost. Since the economic advantage of DH depends on the value of the fuel savings in CHP production compared to individual boiler production, the low value of these savings in Orenberg undermines its economic attractiveness. In addition, the low cost of electricity has led to a low electricity tariff with the result that heat tariffs from CHP plants are high, since they must cover most of the costs related to CHP production.[21] Another example where building boilers is the least-cost option is Sevastopol, Ukraine which has a mild climate and low heat load densities.

Fuel Savings

DE systems allow several types of fuel savings: (a) the absolute amount of fuel used is reduced through increased efficiency when using large HOBs as compared with individual boilers and when using CHP (or co-generation), (b) economies of scale are obtained in fuel purchases in large quantities, and (c) the kind of fuel burned can be switched away from premium or imported fuels to more desirable and lower cost

---

[21] UNDP & ESMAP, 1998.

choices such as refuse or biomass. Where electric chillers are employed, cold storage can allow substantial monetary savings by shifting load away from peak hours.

An example below shows how much energy could be saved by using the same fuel more efficiently in DE systems. Table 4-1 shows the total fuel consumption in two cases: (a) where electricity is produced in a large supercritical condensing steam turbine and heat is produced separately in heat-only-boilers; and (b) where heat and electricity are produced in a large specially designed supercritical steam turbine CHP plant. The fuel savings which can be achieved by CHP operation is 31%.

TABLE 4-1
Fuel Savings for Heat Supply from CHP

	(a) Separate Electricity & Heat Production	(b) CHP	Unit
Net Electric Capacity:	300	300	MW_e
Fuel Input:	773	928	MW_t
Thermal Energy Out from Steam Turbine:	-	460	MW_t
Thermal Energy Out from HOB:	460	-	
Fuel for Boilers @ 80%:	575	-	MW_t
Total Fuel Consumed:	1,348	928	MW_t
Fuel Savings:	-	**31%**	

The fuel savings depend largely on the temperature at which the thermal energy is removed from the steam expansion process in the turbine. Low temperature DH will offer greater savings, and 150°C steam, which is largely used in North American DE systems, or 150°C pressurized water, which is largely used in DH systems in Eastern Europe, will result in smaller savings. The savings also depend on the assumed efficiency of the boilers replaced by CHP. If the boilers are old and operate at high temperature, their efficiency could be as low as 65%, and the fuel savings would then amount to 37%. It should be noted, however, that in order to achieve these savings, the CHP plant must operate in this mode all year. This can only be achieved in a large DH system where CHP plants are based loaded all year. Therefore, annual fuel saving comparisons should be performed.

The cost of fuel, typically ranging from 50-80% of total heat supply costs, is usually the most important cost factor in DH. In the West, fuel costs for large consumers, such as DH enterprises, are usually much lower than for small consumers, such as building owners utilizing mini-boilers, since fuel costs reflect the actual costs of supply. For example, the costs of gas for large and small consumers in Dusseldorf, Germany in 1997 are presented in Figure 4-2 below, with gas prices for small consumers around 2-2.5 times that of gas prices for large consumers. However, today in many Eastern and Central European countries and especially the FSU, gas prices for large and small consumers are not differentiated with small consumers paying about the same as large consumers, with the result in some places that consumers are abandoning DH in favor of building-level gas-fired boilers. Reducing the price distortions in gas and other fuels would have a major impact in favor of DH, since it would be the least-cost long-run heating option in high-density population areas.

Figure 4 -2
Gas Prices for Large and Small Consumers in Dusseldorf, Germany, 1997[22]

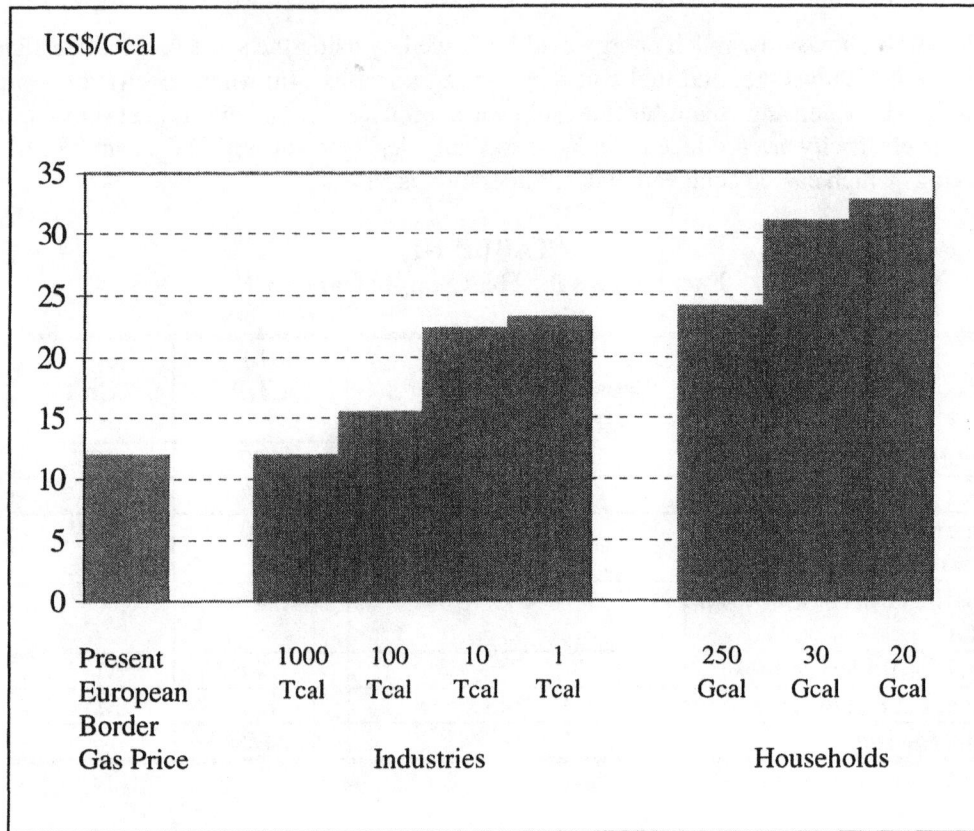

Fuel Flexibility

More economic advantages of using DH can be found in the fuel flexibility that it provides. DH systems with centralized heat production plants make it possible to change into another fuel more flexibly compared to other heating forms, and plants are often designed to burn more than one fuel. Moreover, the assortment of fuels that can be utilized is wider in case of DH as compared to building-level boilers. For example, in addition to premium fuels such as gas or oil, low-grade fuels, such as industrial and community waste that is burned in waste incineration plants, can be utilized as well as bio-fuels, such as waste from wood processing industries and forestry operations. Waste incineration plants provide steam which powers turbine-driven chillers for DC in Indianapolis, Indiana and Albany, New York.

CO_2 emissions can be reduced remarkably with the use of bio-fuels, and opportunities for joint implementation and emissions trade may open up in the near future. Use of bio-fuels creates permanent employment opportunities often in rural areas. Bio-fuels are indigenous and therefore contribute to reducing dependence on energy imports and increasing security of supply. For example, Estonia, which, after independence, converted HOBs in several large cities as well as many cities in the countryside to utilize biomass (primarily wood, wood chips, saw dust, wood waste from forestry operations and peat) was able to increase the share of biomass in its primary energy balance from 3.5% in 1992 to 11% in 1998, thereby significantly reducing energy imports.

---

[22] European Gas Markets, Eurostat – Statistical Office of the European Communities from Riga District Heating Rehabilitation Project, Draft Final Report, Fjarrvarmebyran ab, June 26, 1998.

In addition, geothermal energy can be utilized. In Iceland, 96% of heat is based upon geothermal energy, and in the United States in Western states, there are currently 17 DE systems using a total of 0.61 PJ/year of geothermal energy.

Customer Savings

Since DE systems are generally cost-effective, developers of DE systems are able to offer customers a price that compares favorably with the life-cycle cost of their current systems. However, DE developers and operators are unlikely to offer advantages greater than those they find are sufficient to induce customer conversion, since that would cut into their own profits and therefore limit the opportunities to expand the DE service. Essentially, in a deregulated world, the savings customers will see are those they bargain for and win. Long-term contracts will provide the consumer price stability through insulation from events such as fuel oil price fluctuations. Also, DE systems offer the customer freedom from responsibility and concern for a complex and critical part of their infrastructure, and this "care-free factor," although hard to quantify, offers a real customer benefit.

B. Financial Aspects of District Heating
Tariffs

Western Europe. Generally tariffs in Western Europe are based on the assumption that costs are allocated to customer categories in proportion to their actual costs. Western European DH companies usually combine the market-oriented and cost-based approaches to tariff setting. The goal is that the product's price is adjusted close to, but lower than, the next alternative cost of supplying the particular customer. EU countries will continuously follow EU legislation which requires that competition be taken into account. The key features of a Western European heat tariff are as follows: (a) cost reflective: covers fixed and variable costs separately; (b) market oriented: competitive; (c) motivates energy conservation; (d) simple: easy to understand; and (e) stable: designed for a period of 3-5 years due to the index-linked character with the fuels. It is typical for Western European DH companies to have different individual tariffs based on their costs and market situation. The ratio between the most expensive and the most inexpensive tariff may be about 2:1 in a country. The tariff structure is usually the same for different types of customers of the company, whether industrial, residential or public.

Western European DH utilities have already to a large extent applied a two-tier tariff, which is the most cost-reflective tariff structure currently in use. This tariff structure includes both a fixed (capacity) charge and a variable (energy) charge. The fixed charge can be based on two alternatives: (a) according to the ordered maximum water flow (m^3/h) or (b) according to the maximum heat load of the customer (kW). The same fixed charge is paid every month. The variable charge is based on metered consumption and covers the costs of fuels, purchased energy, spare parts, purchase of external services and profit. The main benefit of the two-tier tariff is that it better reflects the cost structure of DH supply and also takes into account the seasonal and other variations in heat consumption. In addition to the two-tier tariff, a one-time connection fee is usually charged to cover the connection costs of the customer.

Eastern Europe. Tariffs are established for metered consumers on the basis of actual consumption and for unmetered consumers typically on the basis of floor area of living space for heat and number of persons for hot water. The growing tendency now is for Eastern European countries to install meters at the building level in order to based heat bills on actual consumption. Increased metering will lead to higher energy savings for consumers. Metering at the apartment level is generally not possible without further major investments inside the buildings. Heat cost allocators are sometimes used in rehabilitated systems to allocate the building-level heating bill to the individual apartments (mandatory in Bulgaria, for instance).

In Eastern European DH systems, both single-tier and two-tier tariff methodologies are utilized. The most typical methodology is the single-tier tariff where the price of heat consists of one single energy charge and is simple to understand. The single-tier tariff is estimated at the beginning of the heating season but would be insufficient to cover costs if actual heat demand is lower than estimated demand, such as in the case of an unusually warm winter or when consumers invest in regulating equipment which allows them to reduce heat demand. In addition, this tariff does not provide the proper signal to consumers to save energy, because the cost of consuming an additional unit of heat is not transparent. Obstacles to introducing two-tier tariffs include the fact that many municipalities have not yet approved this methodology and also that public awareness raising is necessary before consumers accept the new methodology.

It is not uncommon for a municipality to approve tariffs that do not fully cover costs. In these cases, the municipality usually covers the difference between the accepted tariff level and proposed tariff level from the municipality's budget in the form of an operating subsidy. In addition to the tariff subsidy, the municipality may also provide other subsidies, such as capital subsidies, to DH utilities. Subsidies should be separated from the operations of a DH company, and therefore gradual removal of subsidies is needed.

One particular type of subsidy is the cross-subsidy in the tariff structure between different consumer groups, whereby one group, such as industrial consumers, is charged a tariff higher than its actual cost of supply in order to lower the tariff for another group, such as residential consumers. Cross-subsidies in tariffs are not sustainable because they lead to heat prices for some consumer groups that are higher than the competitive alternative which leads, in turn, to loss of consumers. Elimination of cross-subsidies has already been implemented in several Eastern European countries. In Bulgaria, for instance, the cross-subsidies are planned to be eliminated soon. Improvements are expected in other countries as well. In Ukraine, for example, the heat tariff for residential consumers was established to be 80% of the average tariff in 1998 according to government legislation which meant that the tariffs of other groups, mainly industrial and budgetary consumers, were above the average tariff, resulting in cross-subsidies between the different customer groups. However, in some areas of Ukraine today, the cross-subsidy has been minimized by equalizing the tariff for residential consumers to that of industrial consumers. Moreover, as more Eastern European countries join the EU, the EU legislation will play a decisive role also in tariff matters and in elimination or reduction of cross-subsidies.

CHP Cost Allocation

In the traditional CHP countries such as Germany, Czech Republic, Slovakia and Scandinavia, the electricity market was not open previously for free competition. DH was promoted in order to increase the share of CHP. Recently in Europe, the electricity market has been liberalized adopting market principles, which implies that the accounting of the electricity business must be separated from the total business of an enterprise. Consequently, the need for cost allocation methods for electricity and heat produced in CHP plants has emerged.

For preparation of financial reports of a DH and CHP company and for tax and regulatory purposes, electricity and heat expenses are required to be presented separately. This separation can legally be undertaken according to various methods. Different allocation methods will be adopted by companies in different market conditions, such as companies operating in expanding or saturated markets.

Billing and Collection

In Western Europe, each house, apartment building, industry and other consumer is individually metered, and payment is oftentimes carried out by direct debit of the consumer's bank account by the DH utility, resulting in good payment performance. In Eastern Europe, however, only large consumers, such as industries, are typically metered, and payment performance of non-metered consumers is generally poor. In fact, the main financial problem today facing DH utilities in Eastern and Central Europe and the FSU is the high level of unpaid heat and hot water bills. This problem is highlighted by the fact that registration of a bad debt is not even allowed in Eastern European accounting systems.

A typical feature of billing in FSU countries is the practice by DH utilities of contracting with intermediaries known as municipal house maintenance companies to prepare bills and collect payments for heat and hot water services, along with the payments of other utility bills, from residential consumers. Due to the limited metering of heat production plants and consumption, the costs of heat production and operation are attributed to each building according to estimated consumption in the building, and in the case of apartment buildings, divided among households based on square meters for heat services and on the number of persons for hot water services. These intermediaries also adjust the bills of residential consumers for any discounts that may be applicable under the social assistance programs. Bills for heating and hot water are added to household charges for rent and other utilities for the residential consumers for payment on a monthly basis. While this system of billing provides an incentive for consumers to install heat meters since heat losses are paid by the non-metered consumers, few residential consumers have the ability to pay for the installation of meters.

The high level of unpaid heat and hot water bills by residential consumers is partly due to the billing practices of the intermediaries, which allow residential consumers to pay for their heat and hot water services equally each month throughout the year despite the higher consumption levels in the winter months. In addition, municipalities or state authorities usually fail to transfer all the financial resources to DH utilities to cover the approved discounts for heating and hot water services applied by the intermediaries under social assistance programs. Since DH utilities in this region do not have access to the individual payment records of residential consumers which are maintained by the intermediaries and do not know the level of discounts granted by the local authorities, the utilities are not in a position to identify the portion of the unpaid heat and hot water bills attributable to the local authorities for its overdue social assistance payments and to consumers, and therefore cannot disconnect consumers. Furthermore, as consumers make one monthly payment for all utilities, difficulties arise in identifying what portion of the payment is for heat and hot water when the full monthly payment is not made. As the intermediaries are responsible for the overall maintenance of residential buildings, funds collected for utility bills are also utilized for maintenance and oftentimes the appropriate level of funds is not passed on to the utilities.

To address these difficulties with billing and collection, many DH utilities in Eastern Europe and the FSU are initiating schemes to introduce direct billing to residential consumers after individual building-level meters are installed, which then allows for disconnection of non-paying consumers. The early results have been encouraging in Tallinn, Estonia, for example. Further improvements in billing, collection and monitoring are being pursued by changing from manual procedures of consumer accounts to computerized procedures. In fact, computerized customer information systems, including billing and collection, will be one of the major trends in Eastern European DH systems, and the coverage of computerized customer information systems is expected to increase in Western European DH systems as well. In addition, many DH utilities are now introducing penalties for late payments and discounts for early payments, taking legal actions to recover debts and/or suing for the payment of overdue debts and interest on arrears, and making public announcements in newspapers, for example, of those consumers who don't pay their heat and hot water bills in a timely manner.

For more effective billing and collection in the future, DH companies will need to continue to improve their supply service. By improving the heating and metering service, consumers will be more motivated to pay on time, knowing that they actually received the service and were properly allocated and invoiced. The direct relationship between the DH utility and the residential consumers also necessitates a change in the DH utility's orientation, from one of a DH supply company to one that is concerned with customer service. The improved customer service will mean a fast remedying of emerging faults reported by customers and informing them if the fault cannot be repaired immediately, stating the expected period of delay. This will require special training of staff who have direct contact with customers.

Social Assistance Schemes

In many Eastern European countries, social protection is provided for low-income households in order to mitigate adverse social impacts from increasing prices for district heating and hot water services as well as housing and other utility services, including, electricity, gas, cold water, housing maintenance and garbage removal. In FSU countries, often, housing allowance programs, targeted to poor households, are used to mitigate these adverse social impacts. Housing allowance programs are generally in the form of tied cash transfer schemes according to which social assistance payments are made to providers of housing and utility services, on behalf of beneficiary households.

Other programs to mitigate adverse social impacts from increasing prices for housing and related utility services are cash transfers made directly to needy households (untied cash transfers) or voucher programs. Tied cash transfer programs are often preferred over untied cash transfers because the latter may be diverted for purposes other than the payment of service charges, and cost recovery of utility companies would be adversely affected. Also vouchers provided to low-income households may be diverted for other purposes than payment of service charges, because vouchers can be traded. In addition, voucher programs are difficult to administer.

Housing allowance programs normally have nationwide coverage (as for example in Ukraine and Russia), but are implemented and financed by local governments. These programs are usually means-tested and subsidize household expenditure for housing and communal services exceeding a certain percentage (e.g. 20%) of the income of a household. Similar programs, such as the housing allowance program in Riga, Latvia, try to guarantee a food minimum for households below a certain per capita income threshold. Allowance programs are generally provided within the boundaries of social norms: a limited number of square meters of floor space per person for housing and district heating; a limited number of kilowatt-hours for electricity, cubic meters for gas, and liters per person for water, if the provision of these services is metered. Where utilized, norms for calculating the amount of allowances or social assistance payments per household gives these programs some of the characteristics of flexible lifeline schemes. Further, the system of norms encourages conservation of energy to the extent electricity, gas and water are metered. However, in a number of cases, such as in Poland and Russia, actual expenditures rather than norms were used for the calculation of the subsidy, which reduced the targeting efficiency of the schemes.[23]

Household surveys in Ukraine, Latvia and Russia have shown that targeted housing allowance programs have serious flaws, typically with insufficient coverage of deserving households and substantial leakage to non-eligible households. These flaws can be improved by using additional poverty correlates in determining the eligibility of households for social assistance. Program coverage can be improved through better information dissemination (including public awareness campaigns) regarding the social assistance program. Also, improved inter-agency cooperation, in particular, with municipal housing maintenance organizations, may help increase program coverage.

---

[23] Maintaining Utility Services for the Poor, Policies and Practices in Central and Eastern Europe and the Former Soviet Union, World Bank, March 3, 2000.

On the other hand, sometimes specific rules and regulations prevent eligible households from applying to housing allowance programs. These rules and regulations have to be systematically identified and adjusted. Moreover, the involvement of municipal housing maintenance organizations in the process of billing, collections and data management related to the calculation of subsidized portions of utility bills is often counterproductive and should be addressed. An issue related to housing allowance programs is the frequent lack of financial resources at the local level or the reluctance of municipal governments to comply with their obligation to transfer to municipal utility companies the subsidized shares of utility charges billed to low-income households. Suitable administrative or regulatory mechanisms are required to ensure that local governments transfer to service providers the subsidized shares of charges billed to low-income households.

To upgrade administrative efficiency and effectiveness of social assistance programs, program management, communications and data exchange between social assistance offices, municipal housing maintenance organizations, providers of utility services as well as other government agencies, such as tax authorities, should be computerized. Computerized program management, inter-office communications and operations of data exchange simplify and accelerate processing of applications for social assistance as well as facilitate better follow-up and monitoring.

Financing Instruments

Western Europe. Typically Western European DH utilities finance required investments in their systems through self-financing methods, such as raising equity or internally-generated funds, and through borrowed capital. Other financing instruments which have features of self-financing methods and borrowed capital include bonds with warrants, convertible bonds and quasi-equities. Bonds with warrants include the feature that, in addition to the provision of a loan, the lender has a right to buy a specified number of shares of the company's common stock at some specific price during a designated time period. Convertible bonds include a right of a lender to convert his debt into shares of common stock. Quasi-equities are preferred capital notes which can be qualified as equity and of which interest is paid.

Return on equity can be increased with the use of higher relative shares of debt in a DH company's financing base. The presence of debt, and therefore fewer number of common shares outstanding, results in a higher earnings per share for a company with debt as compared to one wholly financed by equity. However, the leverage includes also a risk that, if the return on the invested capital is lower than the price of the debt capital, the return on equity decreases. In the DH business, especially in many Western European companies, debt is used intensively because the risk is low as a consequence of small variations in demand and hence revenues. Typical ratios of debt to shareholders' equity are between 10:1 and 5:1. Where DH has stiffer competition from alternative heating options, the risks increase, and a debt to equity ratio of 3:2 is considered reasonable.

The risks in connection with the capital intensity of DH may be analyzed from the financiers' point of view. In the case of municipal-owned energy utilities, as in typical in most Western European countries, the municipality usually acts as a guarantor of the loans. In these cases, financial indicators, such as profit margin, operating margin, return on investment, return on assets, quick and current ratios and debt service coverage ratio, are applied. However, the securities given by municipalities to their other creditors will have an impact on the municipality's overall creditworthiness, and thus indirectly on the municipal-owned DH company's loan interest rate.

Eastern Europe. Eastern and Central European and FSU countries have huge unmet investment needs which require external financing, as DH utilities in this region have not been able to build up adequate reserves for future investments through their tariff policies. Investments are usually not needed for increased capacity but mainly for rehabilitation purposes in order to increase efficiency of the existing systems. One of the major problems is to find financing for the investments and to replace the traditional sources of capital funds which were provided from state or local budgets.

Today, the most relevant financing instruments include loans and equity investments of multilateral development banks (MDBs), suppliers' credits and commercial bank loans. The World Bank, European Bank for Reconstruction and Development (EBRD), European Investment Bank (EIB) and Nordic Investment Bank (NIB) have been active in financing the early project investment requirements in the form of loans and/or through acquisition of equity stakes in the DH utilities. Loan financing from these MDBs may be supplemented with risk capital from venture funds. In addition, MDBs also encourage other co-financiers, such as bilateral aid agencies or local or foreign commercial banks, to participate in funding a project. However, in many Eastern European and especially FSU countries, loan financing from local commercial banks may only be available at high interest rates and at short maturities and thus is still not generally available at terms which match the requirements of an investment project. DH investment projects may also be announced under a turnkey arrangement with a successful contractor responsible for obtaining the financing from export credit agencies. The use of suppliers' credits sometimes requires a guarantee from the recipient Government. Generally, a significant part of an investment project can be financed in this manner.

Through the World Bank, the United Nations Development Program and the United Nations Environment Program, countries also have access to financing for certain energy saving projects from the Global Environment Facility (GEF), which was established in 1991 as a pilot program to assist in the protection of the global environment and to promote environmentally sound and sustainable economic development. The GEF operates as a mechanism for providing grant funding to meet the incremental costs of measures to achieve agreed global environmental benefits in the areas of climate change and ozone layer depletion, among others. As of June 1999, 227 projects with a value of US$ 884 million in the area of climate change and another 17 projects with a value of US$ 184 million in the area of ozone depletion have been approved. These funds have helped to mobilize additional resources from other public and private sources.

The parties to the United Nations Framework Convention on Climate Change (UNFCCC) established a pilot phase for Activities Implemented Jointly (AIJ) under the UNFCCC. AIJ implies that governments or companies will contract with parties in another country to implement an activity that reduces greenhouse gas emissions in that country. The World Bank initiated a three-year program on Activities Implemented Jointly (AIJ) in collaboration with the Government of Norway in April 1996. The purpose of AIJ was to gain practical experience through methodological design and information sharing and contributing to the development of a carbon offset mechanism. AIJ will be coming to an end shortly.

Recognizing that global warming will have the greatest impact on its borrowing client countries, more recently in 1999, the World Bank approved the establishment of a Prototype Carbon Fund (PCF), with the objective of mitigating climate change. The PCF pilots the production of emission reductions (ERs) within the framework of Joint Implementation (JI) and the Clean Development Mechanism (CDM). The PCF uses the contributions made by companies and governments to purchase ERs through support to specific projects designed to produce ERs credibly and additional to those that would occur under business-as-usual financing. The procedures to achieve these ERs must be consistent with the emerging regulatory framework of the Kyoto Protocol for the project-based flexibility mechanisms of JI and CDM. Contributors or "participants" in the PCF will receive a pro rata share of the ERs, verified and certified in accordance with agreements reached with the respective countries "hosting" the projects. To be a host country for a PCF project, the country should have ratified the UNFCCC and should be willing to release carbon credits. To-date, the PCF has supported a project in Latvia which has been appraised and is at the stage of negotiating a carbon purchase agreement with the PCF. The PCF intends to collaborate with other purchasers of emissions reductions in the Eastern and Central European region to help catalyze a market in environmentally credible ERs. These currently include the EBRD's Fund managed by Fondolec, and the Dutch government's ERs purchase program.

The financing sources and forms of financing are expected to become more diversified in the coming years. In line with the increasing privatization process in both Eastern and Western European companies, greater investments will be undertaken, which are usually accompanied by transfer of skills and introduction of best utility practice. Moreover, self financing, borrowed capital and bonds are expected to become more established financing instruments, and financing provided from the budgets of state and local authorities will be phased out.

The financial position of existing DH companies in Western Europe has traditionally been strong, which contributes to their ability to borrow in the future. Accompanied with a low risk rate, the DH business has been viewed as an attractive investment objective for many financiers. Eastern European DH utilities are becoming aware of the importance of improving their financial positions in order to raise the necessary capital for investments. Up-to-now, the underdeveloped legal and regulatory frameworks at the state and municipality levels have been obstacles for greater private investor activity in Eastern European countries. When these obstacles are overcome and as the climate for private investment improves, the Eastern European DH utilities will be better prepared to raise the required capital, especially as few companies have any long term debts at all.

5.   Environmental Aspects of District Heating

A. Environmental Benefits of District Heating

DH has a number of environmental advantages compared to individual boilers or apartment stoves. The benefits include: (a) thermal efficiency of large HOBs is usually higher than in small building-level boilers resulting in lower emissions; (b) there exists the possibility to use CHP plants with high fuel efficiencies resulting in lower fuel requirements and lower emissions; (c) there exists the possibility to purchase waste heat from incineration plants and industrial processes; (d) centralized heat production in large plants makes investments for higher stacks and flue-gas cleaning technologies more feasible, thereby enhancing the possibilities for emissions controls in larger systems; (e) better possibilities exist for noise insulation in large units, and (f) leaks and pollution from fuel tanks can be more easily reduced in large units than in a number of small units.

The above mentioned benefits have been realized in many locations and countries. The use of primary energy (fuels) has been significantly reduced due to CHP production and DH use in Sweden, Finland, Denmark, China and Korea, for example. Reduction in fuel use has led to substantial environmental benefits.[24] In 1990, the city of Helsinki received the United Nations Air Award for efficient energy generation. Increased use of DH has been the main factor for improvement in the air quality during the last three decades. In the densely populated cities of China and Korea, abating air pollution has been one of the major reasons to introduce CHP and DH during the 1970's and 1980's. In China, for instance, the pollution from small, coal-fired, individual heating units caused a marked rise in mortality due to respiratory illnesses, which are estimated to be reduced now that DH based on large coal-fired HOBs and CHPs, equipped with flue-gas cleaning, has been and is further being introduced in China's larger cities.

Expected emissions reductions resulting from use of a CHP plant as compared with HOB plants with 65% and 80% efficiencies are described in Table 5-1 below. Because the emissions are specific to the fuel utilized and technology, the plants are assumed to burn #2 oil. As can be seen, CHP provides dramatic reductions in emissions, especially CO_2, for a given supply of thermal energy.

TABLE 5-1
Emissions and Emissions Reductions from CHP

	CHP	Boilers (80%)		Boilers (65%)	
Type of Emissions	Emissions	Emissions	CHP Reduction	Emissions	CHP Reduction
Fuel use -MW$_t$	311	692	55%	852	64%
SO_2	111	377	71%	465	76%
NO_x	21	165	87%	203	90%
VOCs	1.1	1.2	8%	1.5	27%
Particulates	8.8	16	45%	20	56%
CO_2	41	192	79%	236	83%

Emissions are in kg/hour.
VOCs are volatile organic compounds, essentially fuel which has been only partially burned.
Systems provide 554 MW of thermal energy (steam or hot water).
CHP fuel and emissions are incremental above those from power generation alone.

---

[24] Euroheat & Power, 1997.

38

In addition, CHP plants are often equipped with more advanced flue gas cleaning systems than HOBs, including such systems as electric precipitators and desulphurization plants, which would result in even larger reductions of emissions.

Stockholm, Sweden offers further evidence that these calculations are not simply theoretical. As the amount of heat supplied to Stockholm by DE has increased by a factor of ten from 1965 to 1990, SO_2 emissions have been reduced by 95% and particulates by 82%. The reduction is due both to increased efficiency and to cleaner combustion in the larger, centralized boilers. Similar results can be found in many cities with large DH systems, such as Copenhagen, for example.

It is now widely accepted in the scientific community that burning fossil fuels has increased the CO_2 content of the atmosphere and that these increasing amounts of CO_2 and other "greenhouse gases," such as methane and fluoro- and halocarbons, will lead to significant and possibly devastating climate change over the next few decades. Two global conferences have been held in Rio de Janeiro, Brazil in 1992 and in Kyoto, Japan in 1997, with the goal of beginning to address this problem. The "Kyoto Protocol" to the United Nations Framework Convention on Climate Change requires industrialized countries to reduce emissions of greenhouse gasses by an average of about 5% below 1990 levels by 2010. For the United States, the target is a 7% reduction. The Protocol will take force after 55 countries have ratified it. The ratification process is currently ongoing, and by end-September 2000, 84 countries had signed and 30 countries had ratified the Protocol.

Several studies have shown that reductions in greenhouse gas emissions are possible using cost-effective technologies. For example, investments in insulation, heat transfer surface, heat recuperation, electric motor controls and other technologies would be paid for by fuel savings over a 3-10 year period, depending on the technology. However, housing developers typically install fuel-inefficient building heating systems because the capital cost is lower and the developers are not concerned with the recurrent costs of operation. Consumers are usually not sufficiently knowledgeable to anticipate the savings that would accrue. Various forms of government intervention could have a dramatic effect on energy savings. For example, gigawatts of electricity (several full-sized power plants) are not needed today because of the energy efficiency standards for refrigerators enacted in the late 1970's in the United States.

B. Environmental Impacts of District Heating

The environmental impacts of DH arise from energy production and heat distribution. When viewed in terms of the life cycle, the environmental impacts of DH can be divided into: (a) the impacts caused by the construction, use and decommissioning of heat production plant and DH networks, and (b) the impacts caused by the fuel cycle (production, transport, use and disposal). The significance of the different environmental impacts depends, among others, on the scope (local - global), the time scale (temporary - continual) and the severity (reversible - irreversible) of the impact. In general, the main environmental impacts of DH are caused by energy production and the construction and maintenance of DH networks.

The environmental impacts related to heat production include, among others: (a) emissions into the air causing air pollution (leading to public health impacts, acidification and other damage in natural and man-made environments); (b) climate change discharges into the water (e.g. thermal discharges, impacts on aquatic ecosystems); (c) discharges into soil (water and soil pollution caused by such factors as leakage from fuel tanks or chemical containers); (d) use of natural resources (fuel); (e) waste (including by-products); (f) noise and nuisance (e.g. power plants, traffic); (g) visual intrusion (e.g. power plants, boiler houses and above-ground pipes); (h) land-use (use of land for power plants, boiler houses); and (i) occupational hazards (e.g. handling of chemicals, asbestos removal when removing insulation during the rehabilitation of boilers).

The most significant environmental impacts related to heat production plants, including CHP plants, are emissions into the air during energy production. Emissions can be reduced in a number of ways: (a) +improving the thermal efficiency in heat production and networks; (b) installation of flue gas cleaning technologies; (c) fuel switching to cleaner fuels; (d) use of renewable energy sources; and (e) implementation of energy saving programs at the consumer level.

The environmental impacts related to the construction and maintenance of DH networks include, among others: (a) land-use and impacts on ground and plants; (b) use of natural resources (e.g. sand for backfilling the excavations); (c) noise (construction works, pumps, power stations); (d) emissions into the air (overheating of insulation material during welding); (e) discharges into soil (water and soil pollution caused by such factors as accidental fuel or chemical leakage); (f) occupational hazards (e.g. handling of chemicals, asbestos removal when removing insulation during the rehabilitation of DH networks); (g) nuisance (e.g. to traffic); (h) visual intrusion (e.g. above-ground pipes); and (i) waste (e.g. liquid polyurethane components and containers, short pieces of pipes).

The environmental impacts of construction and maintenance of the DH networks are temporary and local and therefore relatively insignificant. Because the DH networks are usually situated in the urban environment, the impacts on ground and plants are minor and recoverable. The DH pipes are mainly installed underground and consequently the land area can continue to be used as before. The materials are used efficiently, usually the waste material is estimated to be left over at less than 1-2% of pipe elements. Also the use of preinsulated pipe elements is common, and asbestos is no longer used as an insulation material. Most of the waste can be recycled, used in energy production or taken to a dump pit. The amount of hazardous waste is small, mainly liquid polyurethane components and containers, which must be treated properly. The noise and other nuisance of construction and maintenance works can be mitigated by planning and scheduling the operations carefully. Also the noise from pumps and power stations can be reduced by technical means.

Solid waste from heat and power production includes ash (largely form coal plants), particulates collected from baghouses and electrostatic and cyclonic precipitators, and sludge from stack gas washing systems. Solid waste constitutes a significant disposal problem, since it is often quite alkaline and may contain heavy metals. Although this material is a disposal problem, it is better to have it concentrated in one place where it can be controlled than expelled into the atmosphere. DE is effective in dealing with solid wastes simply because it is more fuel efficient – less fuel consumed per unit of thermal energy produced means less solid waste to dispose of.

Thermal discharges are a significant environmental impact for condensing power production. DE lessens these discharges by putting the thermal energy to good use. In a backpressure turbine (or combustion turbine or diesel engine) designed to provide thermal energy at a useful temperature and to accept water back at the lowest temperature of use in the thermal application, the reject heat can be reduced to zero at times of peak thermal load. In times of low load, the excess heat will still have to be disposed of, since it is rarely optimal to run a CHP unit to meet thermal load alone. In the case of a condensing turbine with partial intermediate steam extraction, some heat will normally have to be rejected to the environment to match the technical requirements of the turbine, but the amount of heat rejected for this reason will be very small. In the case of CHP power generation, the impact on marine life will be significantly lessened.

C. Western European Environmental Policies

The Western European countries have different national environmental legislation covering energy production. The EU has agreed on a so called "IPPC Directive" which requires that the member countries should implement a uniform procedure when environmental permits are given to energy production facilities, for example. The directive also incorporates the idea of using "Best Available Technology (BAT)." Currently, member countries are changing their national legislation to implement the requirements of the directive.

Countries in Western Europe typically apply a set of environmental and fuel taxes, in addition to VAT. Usually, the taxes are based on the carbon and/or sulphur content of fuels but emission charges may be applied to actual emissions (typically CO_2 and SO_2, but also NO_2 and HCl). The taxes tend to be lower for biomass and natural gas due to their lower CO_2 and SO_2 emissions.

Climate change issues are becoming an ever more important question in the energy production sector. As mentioned previously, the Kyoto Protocol was signed by many of the industrialized countries. All the Western European countries, with the exception of Norway and Switzerland, belong to the EU for which the reduction target for CO_2, CH_4 and N_2O emissions is 8% from the 1990 level to be achieved by the year 2010. The reduction targets vary for different countries within the EU. For Switzerland, the reduction target is also 8%, but Norway is allowed to increase emissions by 1% and Iceland by 10%. The EU has adopted a very positive attitude towards increased use of CHP, and DH as a means of meeting the reduction targets.

D. United States' Environmental Policies

Emissions of traditional pollutants — SO_2, NO_x, VOCs and particulates — are restricted by both state and federal laws in the United States. The federal laws, most notably the Clean Air Act (CAA) of 1970, its amendments in 1977 and 1990, and the implementing regulations of the Environmental Protection Agency (EPA) provide the standards for what levels of air pollution require how much attention and specify various mechanisms by which "non-attainment areas" can attempt to improve their situation.

In one such mechanism, the EPA allowed states or regions to establish "bubbles" within which sources of pollutants (largely SO_2 as an acid rain precursor) were issued "permits to pollute" which decrease in magnitude each year. In what was regarded as a very innovative step, the entities owning the sources are allowed to trade and buy and sell these permits among themselves. In theory, this should lead to a least-cost path to cleaner air. For example, in December of 1997, SO_2 permits were selling for about $100/ton, indicating that technology capable of cleaning SO_2 out of enough exhaust to meet that year's scheduled emissions was available for about that price. This cost can be combined with the SO_2 reductions of Table 5-1 to determine a "value" for CHP with respect to SO_2 emissions, and similar calculations will apply for the other emissions as markets develop for them.

The state regulations implementing the CAA vary considerably, in response both to differing local pollution conditions and to differing political configurations. However, in most states, any new equipment above specified modest sizes must pass through a permitting process, and this would certainly include equipment for any realistic DE system. In many cases in the past, DE systems have not been able to claim credit for the significant emissions reductions they would produce because they could not receive credit for the individual boilers being replaced. In part this is because the large DE system must meet industrial emission standards while the commercial boilers being replaced have essentially no emission standards. In addition, emission credit is often given only if the older boiler is physically destroyed, a step owners are reluctant to take both because it costs money and because the old boiler provides a level of back-up protection for what is to them an untried supply system. Modifications to the regulations to recognize the

41

real lowering of emissions DE makes possible and allow the developers to incorporate those savings into the emissions trading market would greatly enhance the viability of DE systems.

Several studies have shown that such reductions are possible using technologies that are cost-effective. However, objections by interests who fear that they will suffer if strong energy efficiency programs are instituted has led to a stalemate in the United States' Congress and the possibility that the Kyoto Protocol will not be ratified. As shown above, DE could make a significant contribution to this effort.

E. Central and Eastern European and The Former Soviet Union Environmental Situation and Policies

Environmental Situation

In Central and Eastern Europe and the FSU, environmental benefits of DH have been reduced by three key factors. The first is the poor efficiency in the DH networks causing huge losses and increased need for heat production and thus fuel use. The second is the still underused possibility to increase the share of CHP in DH production. The third has been the need to use cheap sources of fuel for heat production in order to maintain competitiveness with individual boilers; coal was preferred to natural gas and heavy fuel oil to light fuel oil. While the disbenefits of utilizing these more polluting fuels could be reduced by appropriate flue gas cleaning technologies, often there has been neither the funds nor the incentive to invest in them. Despite the above described factors that have reduced the benefits, it is most likely that the current environmental problems in Central and Eastern Europe and the FSU would be much more severe if CHP and DH were not used as much as they are.

The measures used for mitigating the environmental impacts of DH are not yet implemented to such an extent as have been done in Western Europe, and as a result, emissions to the air from heat production are much higher than in Western Europe. Also waste management, especially material recycling, is not yet as developed and efficient as in Western Europe.

Environmental Policies

Many Central and Eastern European countries (e.g. Czech Republic, Romania, Slovenia) apply penalties, taxes and other fees on polluters. The penalties are often not sufficient to cover the costs of the needed environmental programs or to control the emissions. Bulgaria applies tax relief for companies which use environmentally sound technologies.[25]

New environmental legislation is beginning to appear, including emissions standards and standards for ambient air quality. Many of the countries are harmonizing their environmental standards with those of the EU. More strict emissions limits are being applied to new power plants, and often also old ones are required to be equipped with modern fuel gas cleaning technologies. Still, many Central and Eastern Europeans live in areas that do not comply with national air quality standards.

Most of the Eastern European countries signed the Kyoto Protocol to the United Nations Framework Convention on Climate Change in 1997. The reduction target from the 1990 level to be achieved by the year 2010 is 8% for Bulgaria, Czech Republic, Latvia, Lithuania, Romania, Slovakia and Slovenia, 6% for Hungary and Poland, 5% for Croatia and no reduction is expected from Russia and Ukraine.

---

[25] US Department of Energy, 1999.

42

6.   Comparison of Eastern European and Western District Heating Systems

A. Extent and Coverage

DH utilization began on a large scale in both Eastern and Western Europe after World War II. Western European countries in the forefront were Germany, Denmark, Iceland, Sweden and Finland. Today, these countries and Austria are the major users of DH. In Eastern Europe, the growth of DH utilization was even more rapid than in Western Europe. Today, DH has over 50% of the market share in Estonia, Russia, Latvia and Lithuania. The DE situation in Europe is summarized in Table 6-1 below. More detailed technical statistics are provided in Annex 6.

TABLE 6-1
District Heating in Europe

Country	Maximum Heat Output Capacity	Heat Delivered to Pipeline System	Electricity Produced from CHP	District Cooling Consumed
	MW	PJ	GWh	TJ
Russia	185,000	2,579	175,870	
Romania	48,206	425	25,762	
Poland	57,210	421	19,770	
Germany*	47,900	352	20,716	
Czech Republic	48,885	237	10,488	
Sweden*	28,050	167	4,507	108
France*	20,519	125	687	1172
Denmark*	15,200	111		
Finland*	16,820	95	10,200	
Hungary	17,800	77	1,635	
Estonia	14,007	46		
Austria*	5,300	36	9,679	
Bulgaria	8,800	30	13,000	
Iceland	1,499	18	107	
Netherlands*	3,979	17	17,805	
Switzerland	2,073	15	296	
Italy*	2,462	10	1,292	117
Slovenia	1,797	9	530	0
United Kingdom*	645	6	678	
Norway	750	5	50	9
Total	**341,902**	**4,779**	**313,072**	

* European Union (EU) members; Belgium, Greece, Spain, Ireland and Portugal are also in the EU but have little or no DE.

Currently, over 50% of Europe's total DH consumption is in Russia. A further 26% is concentrated in the neighboring Eastern European countries. Western Europe accounts for 20%, of which 40% is concentrated in the Nordic Countries.[26]

---

[26] Utility Europe, July 1998.

In North America, DE utilization began earlier around the beginning of the twentieth century but its share of the total heating market is less extensive than in Europe, accounting for only about 4% of space heating and cooling demand. Clearly the systems of Eastern Europe completely dwarf those in the United States.

In Asia, DE utilization was introduced later during the 1970's and 1980's. Its present market share in China, Japan, Korea and other countries with heating/cooling demand is currently about 8-12% and is growing rapidly.

B. Technical Aspects

Older and Newer Systems

In the countries of the FSU and Central and Eastern Europe, DE systems are predominantly based on hot water. These systems tend to be decades old and are prone to corrosion and failure. Especially in the FSU, they have also suffered from the economic and organizational decline that has accompanied the break-up of the Soviet Union. Even before the break-up, efficiency was often low as there was little incentive to minimize losses due to the very low fuel prices. Despite their abundant capacity and sunk costs, these systems are in need of drastic and expensive overhaul. In many ways, they are quite comparable to the older downtown steam systems in the United States, with a similar need for re-capitalization, technical modernization and market-oriented operation.

DE systems in several countries – notably Denmark, Finland, Sweden and the Netherlands - are both chronologically newer and constructed on a more advanced technical model. All use hot water as the distribution medium and are based to a large extent on CHP. All have made successful efforts to diversify their sources of fuels, including biomass (for example, wood chips and straw), refuse incineration and industrial waste heat as well as reject heat from conventional, fossil-fuel power plants. In Denmark, all large refuse incineration plants are connected to DE systems and provide 12% of the total DH energy. As a result of aggressive programs to supply expanding areas around their central cores, all have grown at prodigious rates during the last twenty years. For example, DE capacity in Stockholm, Sweden has grown at nearly 11% per year for the last twenty-five years.

CHP Use

Western and Eastern European countries use both CHPs and HOBs to produce DH. The share of CHP in heat production is, on average, around 50% both in Eastern and Western Europe. The amount of electricity produced by the DH systems running in CHP mode in Europe is shown in Table 6-1 above, column 4. The share of DH systems running in CHP mode varies from country to country, ranging among the EU countries from 22% in France to 92% in the Netherlands. Among the EU countries, 13% of total electrical capacity and 9% of electrical generation is from CHP plants. While CHP is a common feature in these systems, DC is quite uncommon, as European cooling loads are considerably lower than in the United States due to Europe's more northern location.

Fuels Utilized

A wide range of fuels is used throughout Europe as well as in North America. No uniform fuel consumption patterns can be found among Eastern or Western European countries. Use of renewable fuels is somewhat more common in Western Europe than Eastern Europe or North America. Hard coal, lignite and heavy fuel oil are still widely used both in Eastern Europe and the West, although a common trend is the increasing share of natural gas.

44

Major Technical Differences

The major technical differences between Eastern European and Western DH systems are found in the technical configurations, performance and efficiency. The technical characteristics and main problems faced in Eastern and Western European DH schemes are summarized in Table 6-2 below. The description of Eastern European systems refers to traditional systems based on Soviet technology. However, many Eastern European countries are rapidly modernizing their systems in line with Western standards.

Heat production efficiencies and heat and water losses in DH distribution systems differ significantly in Western and Eastern European DH systems. In Western Europe, typical CHP plant efficiencies are around 80-90% and HOB efficiencies are around 90%. In Eastern Europe, typical CHP plant efficiencies are around 70-75% and HOB efficiencies are around 60-80%. Water losses in Eastern European DH systems are typically 5-40 times higher, heat losses 3-5 times higher and the specific heat consumption in buildings up to three times higher than in Western Europe.[27]

The main technical reasons for the poorer performance and efficiency and high emissions in Eastern European schemes include: (a) poor thermal efficiency of HOBs and CHPs mainly due to inadequate operation and maintenance procedures and obsolete automation and control systems; (b) inadequate emission control systems; (c) instead of variable flow and indirect connections, use of constant flow and direct consumer connections which do not allow for efficient operation nor allow consumers to control their own consumption; (d) use of hydraulically separated networks which do not allow for the same flexibility and level of consumer service as the hydraulically interconnected networks used in Western Europe; (e) poor thermal efficiency of DH transmission and distribution networks due to leakages and insufficient insulation of pipes; (f) low usage of metering which leads to excessive consumption and lack of incentives to conserve energy; (g) lack of on-line monitoring systems which does not allow for optimal operation; and (h) lack of preventive maintenance.

TABLE 6-2

Technical Characteristics and Main Problems in Eastern and Western European DH systems

Characteristics	EASTERN EUROPE	WESTERN EUROPE
HEAT PRODUCTION		
Characteristics	- CHPs and HOBs utilized - Undeveloped automation and control systems	- CHPs and HOBs utilized - CHP share varies according to market conditions - Online monitoring and control systems used in larger DH systems
Main Problems	- Low efficiencies - High emissions - Often low share of CHP production	- Further increases of CHP would improve efficiency - Lower emissions but further need to reduce
DH NETWORKS		
Characteristics	- Predominantly constant flow - Low usage of preinsulated pipes	- Variable flow - Preinsulated pipes used widely
Main Problems	- Leakages due to internal and external corrosion - Extensive use of make-up water - Insufficient insulation - Unreliable supply	
CONSUMER		

---

[27] Utility Europe 1998, Power Economics 1998.

Characteristics	EASTERN EUROPE	WESTERN EUROPE
EQUIPMENT		
Characteristics	- Direct (and indirect) consumer connections and sometimes direct domestic water connections - Heat consumption usually not metered - Significant use of large centralized substations serving groups of buildings - Insufficient automatic control	- Indirect (and direct) consumer connections - Heat consumption metered -Automatic control based on outdoor temperature and consumer demand
Main Problems	- High specific consumption - Overheating & underheating	

C. Institutional Aspects

Institutional and Ownership Arrangements

A comparison of institutional arrangements for Eastern and Western European DH systems is shown in Table 6-3. Both Eastern and Western European DH utilities are characterized by predominately public ownership, mainly municipal, although privatization of DH utilities is increasing. In the Western European energy business, DH and electricity activities are mostly concentrated in the same company, while in Eastern Europe, these activities are normally divided into two or more different companies.

TABLE 6-3

Institutional Characteristics in Eastern and Western European District Heating Systems

	EASTERN EUROPE	WESTERN EUROPE
Characteristics		
Ownership	- Municipality, region, state	- Municipality; utility companies
Business structure	- Two or more-company structure - CHP typically in a separate company - Operation of distribution and isolated networks and boilers separately	- DH production, transmission and distribution in same company - DH and electricity often in the same company
Regulation	- Generally treated as a "natural monopoly - Regulated by the public sector - Regulations often included in "Local Public Services Law	- Some times treated as "natural monopolies" and some times as operating in a competitive market - Progress of energy market liberalization has an impact on regulation - Regulation should follow EU legislation.
Type of orientation	- Supply driven	- Demand driven; market oriented

A number of problems arise from the multiple company arrangement and ownership of the DH and CHP systems in Eastern European cities. For one, the sharing of the benefits of the co-generation process requires complex contracts between the CHP plant and the DH network company and is difficult to regulate. The amount of the benefits from the co-generation process, as compared to the production of heat and electricity in independent processes, varies according to the market situation for both heat and electricity, which may move independently. From time to time, the amount of the benefits from co-generation is attributable to the trends in electricity markets, and, from time to time, to the trends in heat market. It is difficult to formulate an equitable benefit sharing arrangement for heat and electricity services that would be able to respond to short and medium-term market fluctuations through contractual arrangements between separate heat and electricity entities.

Additionally, difficulties arise in operating the DH networks and CHP plants in an optimal way. The combined DH system, i.e., DH networks, HOBs and CHP plants, are technically, hydraulically and thermodynamically interconnected. Operating the DH system in an optimal way would require control functions that extend over the whole integrated system. It is difficult to agree on the various parameters and their interrelationships for the control of the whole system in a contractual agreement, especially as the optimization of these parameters would produce different results, both financial and technical, for heat and electricity, if done separately.

Furthermore, it is typically difficult to resolve disputes in daily operation of the DH system. Faults and operation errors on the DH network side may cause physical damage or lost production on the CHP plant side, such as for example, water hammers caused by improper valve operation or a major pipe rupture may cause physical damage at the CHP plant. Similarly, faults and operation errors of the CHP plant may cause financial losses at the network side, such as a CHP turbine trip which would stop the flow of heat from the CHP plant. It is difficult to foresee all possible mishaps in daily operations and agree the dispute resolving mechanism and remedies to the parties in a contractual agreement.

Also, it is also difficult to optimize long-term investments for the DH and CHP system, when approached separately by each company. Investments in DH networks, peaking and reserve boilers and CHP plants should be coordinated to minimize overall operation costs and investment risks and ensure the long-term optimization of the combined DH system, which is difficult to achieve when each company is only considering how to optimize its part of the system. For example, it has been difficult to introduce investments, such as variable flow, which conserve energy, reduce the heat demand and improve the competitiveness of the heating service in a number of FSU countries, because the production plant owner would lose sales and is not necessarily consumer-oriented.

Within a DH system based on hot water as an energy carrier, as opposed to electricity which is easier to dispatch, there are many technical reasons, such as network capacity and hydraulic configurations, that limit the dispatching of heat on economic principles. When a DH system has multiple owners and operators, these reasons can be easily used to discriminate against some of the heat suppliers, resulting in less-than-optimum operation. This may lead to higher prices for heat and loss of consumers to competition. In order to better optimize the DH operations of a system within a city so that DH can compete with other heating alternatives, a number of Eastern European countries have started to merge the various DH companies serving the same area so that heat production, transmission and distribution are contained in the same company, as was the model in the West from the beginning. In other cases where heat production plant, such as CHPs, are maintained in separate enterprises, DH companies are finding ways to create a competitive environment among production plants to bring down the purchase price of bulk heat.

47

Market Orientation

Traditionally, in Eastern Europe and especially in the FSU, the heat consumers have not had any alternative option for heating other than DH. Therefore, the attitudes of the DH companies have been supply-oriented. Further, the bulk heat producer (i.e., large HOB or CHP plant) usually delivered heat to the housing maintenance companies and had little interest in the quality of service and consumer satisfaction, as their sales were guaranteed and they had no direct relationships with small individual consumers.

The situation is now rapidly changing. The majority of countries in Eastern Europe and the FSU have generally adopted legislation that allows the consumers to be able to select such heating option as they wish. Therefore, in principle, DH is no longer protected and is subject to growing competition from other alternatives, especially gas-fired mini-boilers. DH utilities have now adopted a number of strategies to respond to the new market situation. For one, in Russia for example, the owners of DH utilities, such as regional or municipal authorities, try to maintain DH markets through use of regulations such as building permits for mini-boilers, fire codes and building codes to restrict the competition. In the Baltic countries, on the other hand, DH utilities are now trying to genuinely compete with the alternatives by improving the quality of the service (e.g., supplying domestic hot water during the summer months) in order to maintain market share. Many companies are also acknowledging that DH cannot remain competitive with decentralized heating options in all locations and are shutting down their uneconomic DH networks and providing decentralized heating instead.

D. Financial Aspects

The key financial characteristics of Western and Eastern European DH systems are summarized in Table 6-4 below. In Western Europe, DH tariffs are market-oriented and formulated in a way which reflect both the variable and fixed costs. In addition, since tariffs reflect the actual costs of serving different consumer categories, the tariffs encourage energy conservation and are easy to understand. In Eastern Europe, on the other hand, tariffs are usually based on estimated consumption and do not reflect the actual costs of supply. In addition, oftentimes tariffs are established below cost recovery levels, thereby requiring subsidies from their owners, usually the municipalities.

Billing and collection systems also differ significantly in Western and Eastern Europe. Whereas Western European billing and collection systems are typically automated and involve the direct debit of consumers' bank accounts by the DH utility, Eastern European systems involve the billing and collection through intermediaries, which introduces difficulties in establishing direct relationships with consumers and in applying sanctions for non-payment. The result is a generally poor collection performance in Eastern European countries.

The options for financing of needed investments is more limited in Eastern European countries than in Western European countries. Due to poor financial performance and tariffs which have not allowed for the build up of reserves for investments, Eastern European DH utilities have not been able to finance investments from internally-generated funds but have relied more heavily on state and local contributions as well as loans and equity from multilateral development banks and to some extent also on suppliers' credits. Western DH utilities have had a wider range of options to choose from, including self-financing, borrowed capital, and various types of bond offerings. As DH utilities and the respective country's economic performance improves, the range of investment financing options increases.

48

TABLE 6-4

Financial Characteristics in Eastern and Western European District Heating Systems

	EASTERN EUROPE	WESTERN EUROPE
Characteristics		
Tariffs	- Typically single-tier tariffs utilized; - Tariff structure often includes cross-subsidies; - Where tariffs do not cover costs, direct subsidies from municipal owners are often provided;	- Two-tier tariff: cost reflective
Billing	- Billing typically carried out by intermediaries - Based on actual consumption only to limited extent where meters exist - Mostly based on estimated consumption	- Highly automated, service orientation; direct debit of consumer's bank account - Based on actual consumption measured by meters
Collection	- Poor collection rates; - Discounts and social assistance generally underfunded; - Few sanctions for non-payment	- All the bills collected
Financing instruments	- State or local budgets; - Loans from multilateral development banks; - Suppliers' credits.	- Self-financing; - Borrowed capital; - Bonds with warrants, convertible bonds, and quasi-equity.

E. Environmental Aspects

The environmental impacts of the use of DH and CHP are viewed somewhat differently in Western and Eastern European countries. In Western Europe, the increased use of CHP and DH is considered to have enhanced environmental advantages and will continue to in the future. In Eastern Europe, on the other hand, often the use of DH and CHP as well as HOBs is considered to be a major source of emissions to the air. Even though it is true that CHPs and HOBs in Eastern Europe have not been equipped with as efficient pollution controls as in the West, it is also true that if CHP and DH were not used, the environmental problems would be tremendously worse.

Many of the Eastern European countries plan to join the EU in the future and are beginning to harmonize their environmental legislation with that of the EU. Western environmental standards are gradually being adopted. The transition period to the adoption of Western standards will take some time. Even if state-of-the-art flue gas cleaning technologies may be required for new production capacity, the problem of old polluting plants can only be solved at a slower pace.

49

7. District Cooling

A. History and Present State of District Cooling in Europe

Extent and Coverage

In Europe, the first two DC systems were established in the late 1960's in Paris and Hamburg.[28] Since then, a number of schemes have developed. Some of the key installations in Europe are shown in Table 7-1.[29] Most of the systems mentioned in the table are used without DH installations.

Currently, France, Germany and Sweden have the most installations in Western Europe. There are 12 major DC networks in France representing over 450 MW of cooling output capacity. In addition, there are numerous small installations with capacity less than 2 MW_c. Most of the systems are driven by electric chillers, and absorption techniques have very small potential. The main reason is the low price of electricity during summer.[30]

Almost all of the 10 German systems operate with absorption chillers connected to an existing DH scheme. Their cooling capacity ranges between 100 kW and 5 MW. Two important cooling potentials have been identified in Hannover (50 MW_c) and Berlin (30-50 MW_c).[31]

In Sweden, the first system was installed in 1992 in Västerås. By the end of 1997, Sweden had 13 DC networks in operation in the southern and central regions of the country. The total capacity is 137 MW, with annual deliveries of 140 GWh through a distribution network of 35 kilometers.[32]

The largest known unit operating in the Central and Eastern Europe and the FSU is in Tashkent, Uzbekistan. The system serves about 1,000 apartments, offices and others. In Moscow and other Russian cities, there are numerous small installations.

Cooling capacity and length of the networks are very diversified. Small networks may serve just two buildings close to each other and large networks can supply huge business centers, such as La Defénse, Paris, where the cooling capacity is 243 MW.

Technical Aspects

Offices and business premises have traditionally been air-conditioned by using local electromechanical cooling machinery. Older machines often use chloro-flouro-carbons (CFCs) as working media. At the same time, increased use of office machinery, such as computers, has increased excess heat in buildings. Therefore, DC has become a new, increasingly appealing alternative to eliminate surplus heat in buildings.

DC is most advantageous for buildings with high demand for cooling, such as offices, factories, hospital blocks, department stores and other commercial buildings. The greatest potential is in areas with a high density of stores and offices, where the main energy requirement is air conditioning.

---

[28] Westin, 1998.
[29] Delbès and Vadrot, 1997, with data from the Swedish District Heating Association, 1997.
[30] Delbès and Vadrot, 1997.
[31] Delbès and Vadrot, 1997.
[32] Swedish District Heating Association, 1997.

TABLE 7-1
District Cooling Systems in Europe

Country	Sites (no of schemes)	Capacity MW$_c$	Length of pipeline system km	Commissioned year
Austria	Linz	4.2	1.3	1993
Denmark	Herlev			
	Herning	0.9		1997
Finland	Helsinki	1.5		1998
	Jyväskylä			
France	Bordeaux Airport	4.2	1.5	
	Channel, Sangatte	22.9	84.1	
	Lyon	36	10	
	Montpellier (2)	37	13	
	Paris (5)	357	67	
	Villepinte	24	13	
Germany	Bayreuth	3.3		
	Berlin			
	Bremen			
	Chemnitz	6	3.6	
	Dresden	2.9		
	Giessen, Univ.			
	Hamburg	2.7		
	Hannover			
	Kassel	0.5		
	Mannheim	1		
Italy	Bologna, Univ.	2		
	Genova	2		
	Regio Emilla, San Pellegrino	3		
	Vincenza		14.8	
Liechtentein	Liechtenstein	6.5		1973
Monaco	Monaco	12.5	2	
Norway	Baerum	7.5	4	1986
Portugal	Lisbon, Expo 98	22	20	1997
Spain	Barcelona	5.8		
	Valladolid	2		
Sweden	Göteborg	9		
	Lund	9.7	6	
	Norrenergi	15	6.5	
	Stockholm	65	9	1995
	Uppsala	7	3.4	
	Västerås	18		1992
	Other sites (7)	6.6	4.4	
UK	Channel	28.2	109.1	
	Chatham, Kent			
	Heathrow	35		
	Manchester			
Uzbekistan	Tashkent			

DC systems can be constructed as stand alone schemes, but they quite often co-exist with DH systems. The benefits of co-existence lie both in the supply and user side. In production, the use time of CHP plants can be increased in the summertime, and the need for condensing power is thereby reduced. Further, the use of primary fuels and consequently emissions are reduced. User benefits are the moderate investment costs, given the already existing DH network and equipment. The co-existence also enhances synergy in operation and maintenance of the DH and DC networks.

One such type of system involves the production of heat at a CHP plant which is then transmitted through the DH network into the consumers' premises where absorption heat pumps are located. The absorption systems use heat as primary energy to generate cooling energy. During summertime, when the DH supply temperature of the hot water which is empowering the absorption chiller has been reduced from 120 to 70°C, the operation of absorption heat pumps is problematic because the pumps usually require higher operating temperatures. These schemes have been installed, for example, in Helsinki, Finland and Lisbon, Portugal.

In some systems, the problem caused by reduced temperature is solved by isolating some parts of the network and supplying heat at 120°C to the consumers' premises where the absorption heat pumps are located. Simultaneously, the DH network may also provide heat to some consumers. These schemes have been installed, for example, in Seoul, South Korea.

Another type of scheme involves the production of hot water at the CHP plants with the hot water going into the DH networks and the production of chilled water by an absorption chiller located at the CHP plant with the chilled water going into a completely separate DC network. The consumer receives cold water at 6°C and the return temperature is 15-16°C. Consequently, larger pipe dimensions than in DH networks are needed due to a smaller temperature span. These schemes have been installed, for example, in the United Kingdom.

A scheme which does not co-exist with a DH system involves the production of chilled water by electricity-driven compression chillers with the chilled water going into the DC network. These schemes are found in many different locations around Europe. Another example of a scheme which does not co-exist with DH involves the use of deep water source cooling (DWSC), referring to the use of a large renewable body of naturally cold water as a heat sink to produce chilled water as an alternative to using energy-intensive equipment. DWSC is considered as "free" energy. These schemes have been installed, for example, in Stockholm using cold sea water. A high reliability of service can be achieved in such schemes; Stockholm Energi, Sweden, provides its cooling consumers an availability guarantee of 99.7%.

DC systems commonly use chilled water, sent out at 4 to 7°C and returned at 10 to 16°C. The use of ice slurries for cold storage and even for distribution of cooling capacity allows 0°C supply and considerably smaller piping due to the increased temperature drop. The added cost of producing of ice (due to the lower coefficient of performance (COP) of the chillers at lower temperatures) must be weighed against this, and these systems are not yet common.

As explained above, DC and DH/DC systems require chilled water, which can be provided either by central, electrically-powered heat pumps operating as chillers or by central absorption chillers powered either by boilers or by reject heat in a CHP system. Greater efficiency is possible with large central compression chillers since they receive enough maintenance to permit the use of "open" systems where the motor is outside of the refrigerant loop. Smaller chillers appropriate to end-use installation normally have the motor hermetically sealed inside the refrigerant loop with attendant lower efficiency, since the chiller is disposing of the motor heat. High pressure steam DH systems are often used to power absorption chillers at the customer end, but since no cold medium is being circulated, these are considered to be DH systems rather than DH/DC.

Institutional and Financial Aspects

The institutional development of DC normally follows the same ownership form as DH systems in Western Europe. Generally, DC systems are under the ownership of the municipalities and are not separated from the DH businesses.

DC is still a new phenomenon and the investment costs are relatively high. Heat from a CHP plant may be used to generate DC, and in this way, the heat load can be increased outside the heating season, and consequently the utilization rate (i.e. peak utilization time) of the CHP and DH network system rises. This translates into increased revenues from the same assets.

Environmental Aspects

DC has some of the same environmental benefits as DH. The specific environmental advantages of DC include: (a) use of CFCs is reduced due to replacement of old electromechanical equipment using CFC as circulating fluid; (b) "free" cooling resources, such as cold sea areas, can be utilized; and (c) integration of DH and DC technologies reduces use of primary fuels due to reduced power consumption and peak shaving.

B. History and Present State of District Cooling in the United States

Extent and Coverage

The first commercial DC system appeared in Hartford, Connecticut in 1962. Historically, DE systems in the United States did not provide chilled water, since the high pressure steam could be used to operate absorption chillers at the customer's building; these individual building systems were not considered to be DC. This meant that as the DE systems contracted and moved away from co-generation in the 1960's and 1970's, the urban space cooling system became less efficient, since the delivered steam was more expensive. As late as 1992, the only urban systems supplying chilled water to their customers were a selection of the new merchant developers; all of the old urban steam systems were still supplying only steam. Meanwhile, DC systems at independent institutions have been growing steadily over the past few decades and now constitute the bulk of delivered DC in the United States. The recent situation is summarized in Table 7-2. The largest 15 DC systems in the United States are shown in Annex 7.

TABLE 7-2
District Cooling Installations in the United States

	College	Urban & Community	Hospital	Industrial	Military	Other	Total
Number of DC Systems:	1,043	22	1,209	192	107	369	2,943
Capacity (GW):	10.7	1.3	10.9	4.5	21.7	1.2	50.4
Annual Energy(PJ):	61	7	76	22	31	NA	196
Distribution Line Length(km):	1,388	82	336	722	807	200	3,534

Technical Aspects

DC systems are growing in popularity because their large scale makes possible many areas for increased efficiency that are not available in the smaller cooling systems common in individual buildings. Some of these areas are: (a) large compressors with regular maintenance can be of the "open" design, where the motor is not in contact with the refrigerant; this raises the COP substantially above that for "closed" or "hermetic" designs where the refrigerant carries the motor's heat away; (b) large central absorption chillers are more cost-effective than smaller ones and can run on co-generated heat; they are the only way to use reject heat from power plants or industrial processes for building cooling; (c) cold storage (of either ice or cold water) is more cost-effective in large installations and can be used as a load leveling device for electric demand, a matter of interest if there is a significant difference in on-peak and off-peak electric rates; (d) large systems can take advantage of unusual heat sinks, such as ground water, sewage or wet cooling towers, all of which can raise COP substantially; and (e) conversions to efficient, ozone-friendly refrigerants (HFCs, HCFCs, and especially ammonia) is much easier for a large, well-organized enterprise.

Of course, DC also offers building managers the same reliable, very low maintenance services that DH does, since all the complex equipment is managed by the supplier. As a result of this and the cost savings made possible by increased efficiency, developers are finding that new DC systems can compete quite well with conventional alternatives.

Institutional Aspects

The ongoing development of modern merchant DC systems did not occur under the traditional regulatory structure, due to the unwillingness of entrepreneurs to risk capital on a new enterprise when the return on investment was constrained by the regulatory environment. Prior to the development of a climate of "deregulation," DC systems were restricted to institutions such as colleges, hospital complexes, military bases and similar campus-like environments. In this context, the institution could make a judgement on the most cost-effective way to provide for its future cooling needs and develop the system without concerns for rights-of-way and franchise violations. For cooling systems, even the objections of the local utility to CHP did not arise. The results are apparent in Table 7-2, where less than 3% of the installed DC capacity is found in the urban and community systems potentially subject to regulation. In recent years, urban DC systems have also begun to grow rapidly as unregulated service suppliers.

Financial Aspects

DC systems are similar to DH systems in being capital intensive, while offering long-term savings in operating costs. All the considerations of DH apply, with the additional difficulties that come with the fact that DC is even less familiar to investors than is DH and thus faces higher thresholds of credibility. The finances can be helped considerably if the DC system is part of a modern, hot-water-based DE system, since then trenching and equipment costs can be shared.

Integrated system development is not necessary, however, if strong and creative financing is available, as is shown by the solo DC system being developed in downtown Chicago by Unicom Thermal Technologies. While building up a system including massive ice storage that serves 2.3 million square meters of space, this company has acquired the right-of-way to every building served. By including a conduit suitable for fiber optic and other signal cables, this company has constructed an asset—communications access to more than 200 buildings, with considerable added value.

54

C. District Cooling in Asia

Japan is the country leading in DC installations in Asia. In 1993 the number of DC systems was about 120. In the Peoples' Republic of China, there are some installations in Beijing, Shanghai and Zhejiang. In some sites in Korea and the Philippines, DC schemes include 100 km of network pipes and cooling capacities up to 350 MW.[33]

---

[33] Delbès and Vadrot, 1997.

8. The Future of District Heating and District Cooling

A. District Heating Potential

Country Prospects for Growth

Western, Central and Eastern Europe and the FSU. The expectations regarding the development of market share of DH vary significantly country by country both in Eastern and Western Europe. Throughout Eastern and Central Europe, the competition between different heating forms, especially between DH and gas-fired mini-boilers, is becoming more intense as the energy markets are being liberalized. In some countries in Eastern and Central Europe, DH has been extended beyond what is feasible into non-viable areas by administrative decisions, and the market share may therefore decrease. Also, the specific heat consumption of buildings may improve significantly as measures are implemented to improve the energy efficiency of buildings, which may reduce the total heat demand and even the viability of DH in some areas. The potential for the development of the DH market is summarized in Table 8-1 following. The estimates of growth potential are based on the expectations of the national DH associations, which may be optimistic in some cases.

In some countries, DH's market share is expected to grow due to increased use of CHP and increased emphasis on environmental considerations, especially climate change. The EU energy strategy emphasizes the increased use of CHP for which extension of DH use is a prerequisite. The EU's strategy is based on considerations of sustainable development and improving the environment, especially in preventing climate change and increasing the use of renewable energy sources, such as biomass.

Local climatic conditions may initially have been a determining factor for the take up of DH and DC, but they have less impact on the viability of DE today. For example, Sweden is a country with a cold climate, but it is one of the leading countries in Europe in the development of DC systems. Finland is a country with cold climate also, but the specific heat consumption per m^3 in buildings is lower than in many countries more to the south.

The seasonal variation in heat consumption has an impact on the viability of DH. Even in moderate climatic conditions, DH may be feasible if heat consumption does not vary much between different seasons, for example, as in the UK, and/or domestic hot water consumption works as a major component of heat base load. In these cases, the need to invest in significant peak capacity for both production and transmission, which would be used only for a short period of time as happens in countries with cold winters and warm summers, is reduced.

Due to the increasing competition in heating markets in Eastern Europe, new areas are no longer being connected to DH by administrative decisions, as developers are making their decisions based on market considerations. These decisions are not only based on price, but also on future expectations regarding changes in price and image (security of supply and environmental considerations) that have an impact on the decisions. For example, in Sheffield, United Kingdom, DH has been marketed as "Green Heat," because it is not utilizing fossil fuels but rather waste heat from incineration, which has had a positive response by the consumers.

TABLE 8-1
European District Heating Growth Estimates[34]

Country	Prospects for Growth (Current Market Share)	Remarks
Austria	Good (12%)	Expected growth 46% over the 1995 level by the year 2010
Belarus	Limited (50%)	DH has been developed beyond what is economically justified
Bulgaria	n.a. (19%)	DH is competitive with other heating forms
Croatia	Potentially growing (15%)	Decrease in the short term due to decrease in average consumption, speedy growth after the year 2010
Czech Rep.	Potentially growing (33%)	Individual gas heating has been supported by the Government but competition between heating forms is stabilizing as prices increase; environmental considerations will support CHP/DH in the future
Denmark	Good (50%)	Market share expected to grow from 50% to 65% in existing supply areas but not beyond
Estonia	Promising (50%)	Authorities favor DH due to environmental considerations; more intensive competition between heating forms expected
Finland	Good (50%)	Slight increase in market share expected, sales will increase 18% by the year 2010
France	Limited (3.5%)	DH is promoted by national energy policy; low prices of natural gas and electricity make competition difficult
Germany	Limited (large systems), promising (small systems), (12%)	Current growth rate 1% per annum; tough competition; DH has good environmental image
Greece	Good (1%)	New schemes will be established, especially DH together with DC has potential
Hungary	Limited (17%)	Natural gas enjoys cross-subsidies; national energy policy does not emphasize CHP
Iceland	Limited (85%)	Saturated market
Italy	Potentially good (1.5%)	Growth has been rapid, DH enjoys legislative support and financial grants; fierce competition from gas
Latvia	Poor (70%)	Tough competition with local gas heating because consumers want to avoid to pay for the inefficiencies of DH
Lithuania	n.a. (64%)	
Netherlands	Potentially good (3%)	Annual growth in number of connected homes is 1.5% (4% in 1997); average annual growth in sales 3%; according to a consumer survey, half of gas consumers would be willing to change to DH if they had the opportunity
Norway	Limited (3%)	Small schemes based on biomass and heat pumps may be installed but electricity can compete pricewise with heating; in new houses, electric heating is cheaper to install than water-based systems

---

[34] Euroheat & Power, 1999.

57

Country	Prospects for Growth (Current Market Share)	Remarks
Poland	Potentially growing (52%)	Wider introduction of natural gas will bring competition but also benefit DH sector as a cleaner fuel
Portugal	n.a. (~0%)	Limited market for DH due to warm climate, CHP is considered as efficient option to meet demand for heating and especially cooling in special cases
Romania	Potentially growing (31%)	Currently moderate competition from other heating options will become more intense
Russia	n.a. (70%)	
Slovakia	Limited (40%)	DH has relatively weak position in energy policy; strong competition from electricity and gas; market share expected to grow to 45% by 2010
Slovenia	n.a. (9%)	
Spain	n.a. (~0%)	CHP is promoted, possible potential for DC
Sweden	Potentially good (38%)	Competition in the space heating market is strong; CHP is promoted; authorities will support DH in order to reduce CO_2 emissions; DH deliveries to increase by 3% per annum for the next ten years
Switzerland	n.a. (2%)	
Ukraine	Potentially good (65%)	
UK	Potentially good (1%)	Upturn in image; financial support available
Yugoslavia	n.a. (13%)	Intensive construction requires municipal infrastructure

In Western Europe, although electricity consumption is growing, the total market for heating is somewhat stagnant, partly due to improving building standards and insulation. In these conditions, existing DH networks are not likely to grow very much larger. New CHP plants with higher power-to-heat ratios, such as combined cycle plants, might be needed to improve the economic viability of the present DH systems. The future expansion of DH is more likely to shift from existing networks to smaller-scale systems where the distances over which heat is transported are limited. These small systems may involve only a few MW_e in terms of power capacity which might gradually be interconnected as the heat load increases.[35]

United States. The future of DE in the United States is promising. New systems are being proposed, developed and constructed all over the country. Even if no further changes occur in the economic and institutional environment, near-term developments will include: (a) continued upgrading and expansion of existing institutional DE systems in colleges and hospitals and development of new systems; (b) continued development of urban systems by merchant developers, although commonly with continued use of high temperature steam; (c) upgrading of large urban steam systems, such as New York's, even when they continue to be owned by the classic utility, due to competitive pressures; and (d) increases in CHP as the wholesale power market matures and offers generators a level playing field on which to sell their energy and capacity.

---

[35] European Commission, 1997.

Future Technological Development

New technologies and technical improvements will make DH systems more efficient, as discussed below.

DH Production. One major challenge in DH production is to increase its efficiency. One way to achieve this is to increase the amount of heat produced in CHP mode. Another way would be to utilize, if possible, low-cost heat which may be available from industries or incineration plants which may produce such heat that would otherwise be wasted. A third way would be to increase the use of bio-fuels, which are generally available at lower cost than imported fuels.

The efficiency of CHP plants could be improved by utilizing a combined cycle, gas-fired process which would allow heat to be produced almost at zero cost if the co-generation benefits are allocated to heat production and electricity is sold at electricity system long-run marginal cost. In Western Europe, coal has been the dominant fuel for these plants, but its use has decreased slightly whereas the use of natural gas has increased dramatically as more gas is being used for CHP production, as shown in Figure 8-1 below. Natural gas is the cleanest fuel when compared with other fuels and contributes substantially to the solution of several environmental problems (climate change, local air quality problems, acidification). However, until recently, there have been several factors that have slowed down the growth of gas use. The reasons were mainly political; it was only in 1990 that the EU invalidated the directive that prohibited the use of gas for electricity production, and consequently CHP production. The prognoses of three major institutions, International Energy Agency, the United States' Department of Energy and the EU, on the development of natural gas use are surprisingly similar; the use will increase faster than that of any other fuel, by 2.7% per year. This is a general estimate and is not limited to the use of natural gas for DH production.

FIGURE 8-1
Primary Energy Input for DH Production in the EU, 1994-97 [36]

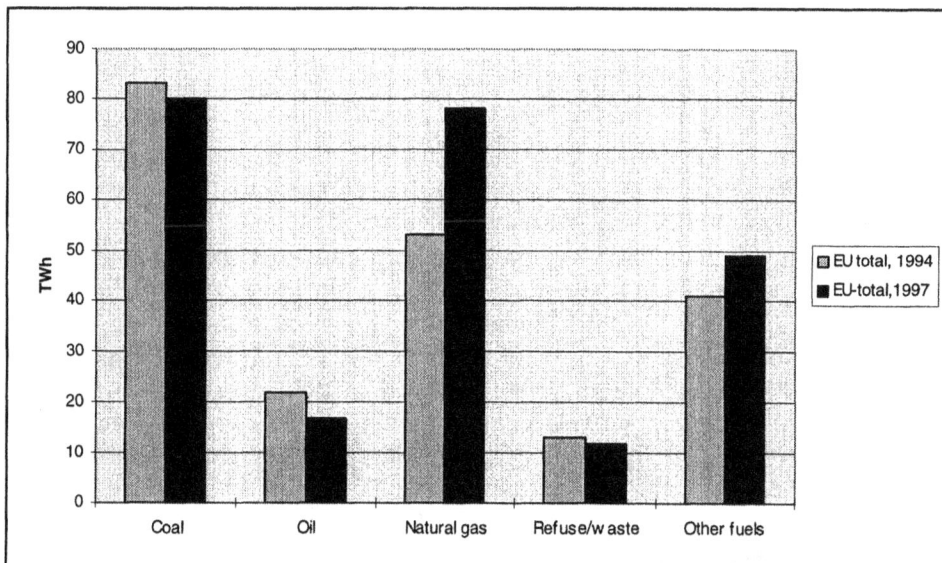

[36] Euroheat & Power, 1999.

In Western Europe, the share of waste and refuse in heat production has been reducing in the recent past, partly due to the new more stringent emission limits established for waste incineration, although the share may increase in the future. Recently, decisions made in the EU require that waste should be recycled and used for energy production and that dumping the waste to landfills should be minimized. Constructing waste incineration plants close to residential areas, which would be a prerequisite for DH production, may raise hard resistance though. In cases where they are constructed, if equipped with modern emission control technologies, they are likely to cause less environmental damage than individual heating or DH based on old coal or oil-fired boilers.

The use of bio-fuels, such as peat and wood waste, is slowly increasing. The trend is likely to continue partly because the use of bio-fuels can contribute towards meeting the Kyoto requirements and partly because significant emphasis is being put on research and development which is likely to improve the competitiveness of bio-fuels. Bio-fuels are usually fuels available locally, and the fuel supply often has a positive impact on local employment. DH can be utilized in decentralized small units which match well with bio-fuel harnessing.

In Eastern and Central Europe and the FSU, natural gas and coal are the most commonly used fuels for DH production also. The trend in this region is also moving towards increased use of natural gas. The use of inexpensive fuels, such as mazut, low grade coal and oil shale, will continue. The problems related to inefficiency and high emissions can be reduced, in addition to use of efficient emission control technologies, by increasing the share of CHP production and reducing the number of HOBs.

Higher utilization of investments in heat production plants can be achieved also by increasing the use of DC together with DH.

Heat Storages. Heat storages can be used to supply heat during peak load hours, by storing heat in off-peak hours in heat storage tanks, thereby reducing the investment requirements for peak capacity. Storages also work as reserve capacity in planned and unplanned interruptions of supply, among other things. Heat transfer networks have traditionally been used as short-term heat storages (for a few hours), but if additional storage capacity is needed, then heat storages are used. Heat can be stored by changing the temperature of storage media, as in a water tank. The new technologies may be based on chemical reaction energy or phase changes (steam accumulators, different salines, paraffin and fatty acids). Also, seasonal storages have been used at an experimental level but their viability has been questionable.

DH Distribution. One of the future trends for improving the efficiency of DH is to decrease the flow temperatures in the distribution networks, lower than the previous standards, and to increase the cooling of the distribution medium (hot water) by the consumer. This improves the efficiency by (a) increasing the electricity yield per each unit of heat produced in CHP, (b) reducing the pipe sizes and pumping energy needed, and (c) reducing the heat losses due to the lower temperatures. In Western Europe, the maximum supply flow temperature is currently 115-120°C, and the trend is to reduce it to 90-100°C. The standard cooling is 50°C, but in some systems a level of 70°C has been achieved. Consequently, the return temperature would be 30-35°C. In Eastern Europe, the maximum supply flow temperature has traditionally been 150°C, and the trend is also to reduce this as much as possible. At present, it is rare to find temperatures more than 130°C being used. In the Eastern European systems, the opportunities for efficiency gains through reduced temperatures and improved cooling are substantial.

In Eastern and Central Europe and the FSU, one of the challenges is to convert DH schemes from constant to variable flow operation, which improves the overall efficiency in several ways. Variable flow allows for automatic temperature control at the building level, which has been shown to reduce the specific consumption of buildings by 15-25% when compared to constant flow. Variable flow also reduces the pumping energy requirements, typically by two-thirds when applied to existing constant flow system.

Further, it allows dispatching of the heat load to lowest cost plants, typically increasing the capacity utilization of existing CHP plants by 20-50%. In addition, it allows the use of lower supply temperatures and increased cooling, thus reducing heat losses of the network.

Also important for improving efficiency of Eastern and Central European and FSU DH systems is the elimination of large central heat substations and adjacent four-pipe secondary networks, which are sources of large heat and water losses and severe corrosion requiring frequent replacement of equipment and piping. Additionally, heat metering needs to be installed. Improvements in metering technology have been taking place and are continuing, replacing meters with mechanical moving parts with, for example, ultrasonic and magnetic meters where there are no moving mechanical parts. The new technology has reduced, and will further reduce, the cost of metering and the need for maintenance. The new meter technology is also very reliable.

Other efficiency gains in Eastern and Central Europe and the FSU will be possible through the increased application of indirect consumer connections instead of direct connections. New technology is making heat exchangers, that are the essential component of indirect connections, inexpensive.

Western countries have been using preinsulated DH pipes since the 1970's, and these pipes can be considered traditional technology in the West. However, in Eastern Europe, these pipes have been available only recently and therefore there is a tremendous potential for rehabilitating the leaky parts of the Eastern European pipe networks, of which there are many, by using preinsulated pipes. Preinsulated pipes are a bulky item and therefore are expensive to transport over long distances. The local preinsulated pipe industry has already developed in some more advanced Eastern European countries, such as Poland and Czech Republic, and is developing quickly in FSU countries. Lenenergo in St. Petersburg, Russia has been producing preinsulated piping based on Western technology since 1996.

New piping materials are being developed. For low operating pressures (<6 bar) and temperatures (<90°C) and small pipe dimensions, pipes made out of plastic, such as polyprophylene or polybutylene, or composite materials may replace steel pipes.

Specific Consumption. Specific consumption is expected to decrease throughout Europe and North America for several reasons, including improved insulation and other changes in construction, better possibilities for consumers to control their consumption, increased awareness due to information dissemination, and greater access to financing instruments for promoting energy efficiency. Improved ventilation in buildings, however, may slow down the decrease. After the oil crises, the applied construction technologies emphasized more energy efficiency than proper ventilation, which contributed, in some cases, to health problems related to mold and mildew, among other things, which now are being addressed.

Operation and Maintenance. The improvement of the operation and maintenance practices in Western and Eastern DH systems is one key area for increasing DH service levels and reducing the cost of the service. Preventive maintenance practices, combined with modern computer-based operation and maintenance management systems, will enable heating companies to improve their performance and, in cases of breakdowns, to act faster and shorten the disruptions in heat delivery. With the introduction of preventive maintenance activities in Eastern European DH systems, large improvements can be gained by reducing heat and water losses, extending life time of pipes and equipment and, especially, improving safety of the systems.

Distributed vs. Centralized Systems. While this report is focussed on centralized DE systems where thermal energy is produced at one location and shipped out to an extended customer area by piping, recent technical developments may make a completely different option practical, i.e., the distribution of small

CHP units among buildings, sized to meet the thermal load of the building where they are located. This would eliminate the need for thermal transmission and distribution systems, although auxiliary boilers might be required in some or all buildings. Alternatively, these in-situ sources could be coupled with thermal transmission and distribution systems to provide load balance and control.

The two technologies making this possible are, first, increasingly intelligent controls for small diesel or gas piston engines, allowing them to run largely unattended. Such systems are available today, although reliability and lifetime costs are yet to be proven. Second, the impending availability of "microturbines," tiny combustion turbines based on automotive supercharger technology and using air bearings and high-speed alternators has aroused considerable interest. Having only one moving part, they are expected to have very high levels of reliability and, when coupled with suitable heat recovery systems, should prove efficient as well. Microturbine manufacturers claim they will become commercially available around the year 2000.

Impacts of Energy Policy

The national energy legislation in each EU country will continue to follow the corresponding EU legislation and international agreements. The energy policy of individual EU states will concentrate on the energy economy as a whole, different energy sources and forms as well as their mutual shares, without attempting to regulate in detail the construction of separate power plants, as before. In most Western European countries, the energy policy measures will focus on the following: (a) global greenhouse gas emissions reduction by 5% from 1990 levels by the year 2010; (b) promotion of an energy production structure towards one with an energy balance utilizing less coal; (c) promotion of bio-energy; (d) promotion of other renewable energy sources such as wind and solar; (e) flexible and least-cost energy supply; (f) promotion of energy market liberalization; (g) integration of separate national markets; (h) promotion of efficient use of energy and energy savings; (i) promotion of high level energy technology; and (j) security of energy supply and promotion of energy from indigenous sources.

The goals for CO_2 emissions will play an increasing role in Western European energy policies and will lead to a larger utilization of CHP to reduce fossil fuel consumption. As a general guideline, in many Western European countries will be the requirement for increased use of natural gas, bio-energy and CHP in order to fulfill their commitments in terms of decreasing greenhouse gas emissions stipulated in international agreements.

The political decisions on closing nuclear power plants in some Western European countries, if implemented, will have a significant impact on the demand for new CHP capacity construction. Closing nuclear power plants causes the need to produce the same electricity in other power plants, and therefore it is important that this new generation takes place in plants, such as CHPs, that comply with the policy goals to produce the minimum amount of CO_2 emissions. In Eastern and Central Europe and the FSU, the possible closing of nuclear power plants would create an even larger potential for CHP plants.

In line with power market liberalization, as old power plants in Europe are retired, the new power plants are most likely to be CHPs, which are more competitive than alone-standing condensing power plants. The differences in the development potential of new CHP plants between the countries in Europe are large and depend on the available heat markets. In Western Europe, the UK, Italy, Turkey, Spain and Portugal offer the best potential, while Poland, Hungary and the Czech Republic will lead the development in Central Europe. The liberalized power markets in Finland and Sweden are also likely to result in some new CHP capacity in these countries, although the DH market is already largely saturated.

Institutional Aspects

Privatization. A clear trend in DH ownership is the change from a wholly or partly publicly-owned DH enterprise towards a more private one. This is occurring in both Eastern and Western Europe. With privatization, DH enterprises have access to new sources of financing for necessary investments that improve the efficiency of the DH service. In addition, privatization usually leads to more effective operations through the streamlining of the organization, allowing DH to improve its competitiveness with other heating options.

Outsourcing. Outsourcing in DH companies is expected to increase both in Eastern and Western Europe. This is a natural consequence of the tendency towards greater cost efficiency and the DH companies' willingness to concentrate on their core competence areas. In the case of a DH company, activities typically outsourced include transportation and maintenance functions and some specified tasks in administration.

Environmental Aspects

Driving Factors. Environmental aspects are becoming ever more important in making decisions in the energy sector. There is a growing awareness of the impacts of air pollution on health and the risks of climate change. Climate change is feared to have a significantly deteriorating impact on living conditions. The World Health Organization (WHO) has estimated that, on average, every person in Europe loses one year of life due to air pollution.

A number of driving factors related to environmental issues will have a significant impact on the market share and development of CHP/DH systems in the future. For one, consumers are becoming more conscious of environmental issues, and environmental arguments are commonly used in marketing in the energy sector. The pressure from different interest groups, such as customers, owners, investors and authorities, has led to companies implementing voluntary programs and supporting management systems in order to improve the level of environmental performance even above legislative requirements.

International agreements of which possibly the most important is the Kyoto Protocol on Climate Change will also have an impact on the future of CHP/DH. Increased use of CHP in connection with DH is one of the main alternatives to increase the efficiency of fuel use and reduce greenhouse gas emissions.

In addition, air quality guidelines are becoming more stringent both in Eastern and Western Europe and will also have an important impact on the future development of DH. Today, ambient air quality does not meet the applicable guidelines in all countries, especially in Eastern Europe. One of the most efficient ways to improve air quality quickly is to reduce the use of small boilers in buildings and to switch to DH.

Green Electricity and Heat. The growing environmental awareness among consumers and the voluntary environmental programs of industrial companies have increased requirements for more environmentally sustainable energy production, such as DH and CHP. In Nordic countries, environmental criteria or labelling for environmentally sustainable energy products has been implemented in co-operation among energy producers, authorities and environmental organizations. The environmental arguments are commonly used in marketing eco-electricity (i.e., green electricity) and eco-heat (i.e., green heat).

Environmental Management. Several power and heat producers, especially in the DH and CHP area where better performance can be shown in the environmental field, are implementing voluntary environmental policies, programs and supporting management systems in order to improve the level of environmental performance. Most of the companies are now making an effort to gain environmental certificates (ISO 14001 and/or EU's EMAS) for their environmental management systems. The first DH utility in the Nordic countries to receive an environmental certificate in accordance with ISO 14001 was Kalmar Energy AB from Sweden in January, 1998.[37]

Since 1996, all DH utilities in Denmark with a capacity of over 50 MW have been under an obligation to adopt green accounting practices. Green accounting must include the use of energy, water and raw materials as well as the quantity of polluting agents, emissions and refuse.[38]

Several companies are publishing specific environmental reports with their annual reports. The objective is to provide information on the environmental performance of the company to the interested parties. A public Environmental Statement is a requirement of the EMAS. This environmental report has to be validated by an external accredited auditor.

Eco-efficiency is a new concept describing the use of natural resources in relation to the produced unit of energy. Environmentally sustainable development can only be achieved by "producing more from less." Specific indicators describing eco-efficiency are being developed for several industrial sectors. In the future, the authorities and investors will probably use these indicators in order to assess the economic and environmental performance of the companies.

Environmental Costs/External Costs. Environmental costs are the costs of environmental damage on human health and man-made and natural environment. If environmental costs are not paid by the polluter (e.g., municipality pays for health impacts but does not receive compensation from polluters), they are called external costs. Internalization of environmental costs means, in practice, making the polluter pay. Internalization of external costs may have a significant impact on the competitiveness of fuels, if environmental taxes are levied based on the damage they cause, and consequently on the cost of heat and electricity produced from these fuels. When internalization of external costs are considered, DH and CHP fare especially well in comparison to other alternatives to heat and electricity production.

In the context of EU's Fifth Environmental Action Programme, the internalization of external costs and benefits in the energy sector through tax incentives is a key priority for the integration of the environment into other Community policy areas.[39] Significant interest in these principles exists and the practical applications are being discussed in different countries, also outside the EU. However, significant uncertainties, especially regarding the costs involved with climate change, slow down practical implementation.

In Norway, energy price and tax policy applicable to fuel usage allowing for a reflection of the environmental costs is under preparation. The Swiss Association of District Heat Producers and Distributors is in favor of internalizing the external costs to reflect the actual costs of each form of energy use and charging the consumers correspondingly.[40]

B. District Cooling Potential

Prospects for Growth

Though the current use of DC is rather limited in Europe, the application of DC is increasing quite rapidly. The leading European country, in terms of installed capacity, is France followed by Sweden, the United Kingdom, Germany, the Netherlands and Portugal.

---

[37] Gunnarsdottir, 1998.
[38] Gunnarsdottir, 1998.
[39] European Commission, 1997.
[40] Euroheat & Power, 1997.

The growth potential for DC is significant. For climatic reasons, the need for cooling is huge in Southern Europe. The cooling needs are growing also in countries with colder climates. The heat loads in offices have increased due to the growing number of office machines, and consumers expect air conditioning in commercial facilities during the warm summer months. The existing DH networks and CHP plants enable rather easy practical adaptation of DC technology.

Key Aspects Affecting Future Development. The technology behind DC is rather new for Europe and it is often not considered to be economic for wide application yet. Large-scale DC systems face high capital costs for energy transfer infrastructure. The efficiency of existing technologies and the high investment costs are limiting factors for the development of the DC concept today. There is a need to decrease the investment costs of the systems. A lot of emphasis is being put on research, and economies of scale in equipment production are likely to reduce the costs.

In general, cooling is not a necessity like, for example, water supply and sewerage or heating in cold climatic conditions. Typically, cooling is being introduced in pace with the overall improvement of living standards. However, this does not mean that cooling would be a mere comfort factor. It is a necessity for cold storages, shops selling foodstuffs and in hospitals. It also improves productivity in offices and other places.

DC is very site specific. The relatively high investment costs and locally unique customer situations, in combination with the operator's background, lead to site specific choices of technology and strategies. DC is also subject to a high degree of uncertainty related to the difficulties in judging the potential market, the growth rate of the system, future competition from substitutes and future use of energy.

However, there are positive practical examples of start-ups of DC systems on a commercial basis. Many DC systems have grown larger than first expected, which implies that DC is a competitive idea and service. There are numerous installations in France, and the development continues. In Sweden, a country with a cold climate, several schemes have been established and the good experience so far has encouraged further plans.

Driving Forces. The driving forces for DC vary by country and include the following: (a) high quality office environment demands; (b) climate (hot climate in Malaysia and southern United States; warm summers and cold winters in Sweden and northern United States); (c) energy situation (few indigenous energy resources in Japan; energy efficiency concerns in Europe); (d) city planning (city planning and city planners have affected, for example, the schemes of La Défense, Climespace, Unicom and Stockholm Energi); (e) facility management, outsourcing (building owners tend to look at strategies to avoid heavy and complicated investments, and operations of complicated technical machinery; DC also frees space and gives knowledge about costs); (f) environmental factors (CFC abolishment leads to large exchange of cooling equipment at the same time; DC is considered as "greener" technology); and (g) other institutional factors (legislation, interest rates and other stability factors are important).[41]

Technical Development

Technology is being developed to achieve better cooling of the DH supply hot water that empowers the absorption chillers to allow for lower input temperatures. In single phase absorption chillers, the cooling of the DH water is currently only 10-15°C, and in double phase chillers 25°C (if the input temperature is 80°C). The target level is 50-70°C cooling of hot water, similar as for DH operation.

---

[41] Westin, 1998.

Thermal energy storage (cool storage/hot water storage) may be an important strategy for optimizing DH and/or DC systems. In general, storages allow a fairly important reduction of investment and operation costs, while enabling a more even operation of the energy production equipment. Hot water storages already are an established technology, but cool storages and the improvement of storage capacity in hot water storages still require further research.

9. The World Bank's Role in the District Heating Sector in Central and Eastern Europe and the Former Soviet Union

A. Macroeconomic and Fiscal Impacts of District Heating

Energy Wastage

Today, in many countries of Eastern and Central Europe and the FSU, energy waste from DH systems is high due to years of lack of maintenance and technical upgrading. For a long time, low cost energy supplies from Russia allowed these countries to postpone the introduction of modern energy efficiency technologies in supply and demand as well as the process of commercialization and restructuring which was taking place in Western countries that were facing much higher international energy prices. Present heat demand per square meter of building in this region is approximately 40-100% higher than in Western Europe, and energy losses in DH transmission and distribution are in the range of 15-35% as compared to less than 10% in Western Europe.[42] There is a substantial potential for improving technical standards and efficiency in the DH sector.

Rapidly Increasing Fuel Import Prices

Fuel prices, particularly import prices, have increased dramatically for Eastern and Central European and FSU countries since 1991, the year of the breakup of the Soviet Union and the collapse of the Council for Economic Assistance (CMEA) system, which led to a requirement for payment in hard currencies instead of barter. Fuel prices quickly approximated world market levels. The higher fuel prices were translated into higher prices for heat and electricity services. The rapid growth in fuel and energy prices after 1991 is illustrated in Table 9-1 below for the case of Estonia, which is typical of many FSU countries. Particularly striking was the growth in heat prices in Estonia which increased about 60-fold alone during 1992. Household incomes, on the other hand, grew at only about 5-fold during the same period.[43] It was not uncommon in energy importing countries for heat bills to represent 60-90% of total monthly household income and in some cases even exceed monthly household income during the early years of independence. As a result, consumers were not able to pay their heat bills in a timely manner.

TABLE 9-1
Growth in Fuel and Energy Prices in Estonia After 1991[44]

Fuels and Energy Services	Unit	Prices			
		Before 1991	January 1991	April 1992	April 1993
Oil shale (domestic)	EEK/t	0.42	1.67	24.70	36.00
Coal	EEK/t	4.25	6.70	411.00	405.20
Heavy fuel oil (HFO)	EEK/t	3.05	8.00	825.00	1,381.00
Natural gas	EEK/1,000 m3	1.85	6.10	539.00	1,321.60
Diesel oil	EEK/t	13.90	n.a.	1,590.80	3,309.00
Gasoline	EEK/t	26.00	n.a.	3,144.80	3,927.00
Heat from HFO	EEK/Gcal	1.00	2.00	120.00	220.00
Electricity	EEK/MWh	2.30	5.00	90.00	150.00
1 US$ = 12-13 EEK					

---

[42] ESMAP, 1998.
[43] World Bank Staff Appraisal Report, Estonia District Heating Rehabilitation Project, May 5, 1994.
[44] Estonian Institute of Energy Research, 1994.

Introduction of Subsidy Schemes and Impacts on Government Budgets

In response, many Governments introduced subsidy schemes to heat producers and/or to consumers, but the level of subsidies was generally inadequate to cover all the costs of heat supply and provide a sufficient level of assistance for those households requiring support. Consequently, consumer debts to heat producers grew substantially, weakening the financial position of the utilities. While the situation has improved somewhat since the early 1990's, still, in many countries, consumer debts to DH utilities remain high. Although prices of energy inputs to DH have stabilized and now reflect near world market levels in most countries, the requirements for subsidies to households are expected to continue over the next 5-10 years, until household incomes grow commensurately with the rise in service prices. Expenditures for subsidies have therefore become a major drain on government budgets.

TABLE 9-2
Value of Energy Imports in Eastern and Central Europe and the FSU, 1998

Country	Value of Energy Imports (US$ million)	% of GDP	% of Total Imports
Albania	86	3	10
Armenia	220	12	27
Azerbaijan	63	2	4
Bulgaria	1,151	9	23
Croatia	818	4	9
Czech Republic	1,867	3	6
Estonia	305	6	8
Georgia	186	4	16
Hungary	1,390	3	6
Kazakhstan	618	3	9
Kyrgyz Republic	220	13	29
Latvia	116	2	4
Lithuania	770	7	14
Macedonia	163	5	9
Moldova	245	15	23
Poland	2,964	2	6
Romania	1,687	4	14
Russian Federation	1,654	1	3
Slovak Republic	1,419	7	11
Slovenia	590	3	6
Tajikistan	271	21	35
Ukraine	6,170	14	38
Uzbekistan	16	nil	1

Dependency on High-Cost Fuel Imports

Many countries in Eastern and Central Europe and the FSU are heavily dependent on high-cost imported fuels for their DH systems and other energy and industrial activities. The values of fuel imports in 1998, where data is available, by country are shown in Table 9-2 below.[45] The high level of energy imports created a demand for foreign exchange that most economies were unable to meet. The difference has been

---

[45] 1998 Country at a Glance and 2000 World Development Indicators, World Bank.

financed in some countries by payment arrears and balance-of-payment support from the IMF and World Bank. Reducing wastage in DH systems would help to reduce the need for high-cost energy imports. In this regard, DH with CHP is the only way to combine production of heat and electricity at the level of 90% efficiency in thermal power plants, leading to reduced fuel requirements. Smaller DH systems with HOBs can also achieve high levels of energy savings, also reducing fuel requirements. Even if DH is scaled back in Eastern Europe, the remaining viable core needs support and rehabilitation and can contribute to reducing fuel inputs. DH systems already exist in the countries of Eastern and Central Europe and the FSU and offer a good opportunity for reducing high-cost imported fuels.

B. Social Impacts of District Heating

Households Connected to DH

DH systems are predominantly an urban phenomenon connected with high population density, and poverty in urban areas is generally lower than in rural areas. These factors help to explain why the shares of households connected to DH in Eastern and Central Europe and the FSU, when viewed from a national perspective, are considerably less than 100%, and why the share of poor households connected to DH are generally underrepresented. Table 9-3 below shows the results of an analysis of surveys in 8 countries to determine the proportion of households connected to DH and the share of poor among households connected to DH. The table shows that Ukraine and Armenia are exceptions to the general trend that poor households connected to DH are underrepresented.

TABLE 9-3
Households Connected to District Heating [46]

	Poverty Group	Armenia 1996[47]	Croatia 1998	Hungary 1997	Kyrgyz 1999	Latvia 1997	Moldova 1998	Russia 1996	Ukraine 1996
% of Households Connected	Non-poor	9.0	33.4	26.6	30.0	69.9	35.9	72.7	31.2
	Poor	10.4	7.8	14.8	12.5	49.0	23.1	62.5	36.9
% of Poor Among Households Connected		33.1	5.2	10.1	11.5	13.9	15.9	27.3	29.9

It may also be suitable to assess, at the level of individual cities, the extent to which the poor are represented among households connected to DH. At this level, the poor may not be underrepresented. Often, 80-90% or more of the population of a city are connected. Those not connected usually belong to the very rich living in new individual houses. However, it has been very difficult to obtain from the cities reliable figures on the share of households connected to DH as well as the type and number of households not connected.

Affordability of DH

Affordability of DH by households remains difficult today, but affordability of DH by poor households[48] remains particularly difficult. While the situation has improved since the early 1990's, nevertheless by

---

[46] Maintaining Utility Services for the Poor, Policies and Practices in Central and Eastern Europe and the Former Soviet Union, World Bank , May 1, 2000.
[47] Households with connections to non-functioning utility services are not considered connected.

1996, in such countries as Russia, Estonia, Lithuania and Ukraine, the cost of heat and hot water supply for a typical apartment still represented a rather high level of about 20-40% of the disposable income of an average household. The share of household income devoted to heat and hot water services by poor households was even higher. Often households in the bottom per capita expenditure quintile spend 4-5 times the share of expenditures that households in the top quintile spend for DH and hot water, as has been shown in recent surveys in Latvia and Ukraine. These figures can be compared with the figure of no more than 8% of average household expenditures which telecommunications, water and domestic energy represent in EU countries.[49] Expenditures for heat and hot water represent still today, in many cases, the largest single expenditure in household budgets.

Poor households, i.e., those households in the bottom expenditure quintile, use most of their expenditures for food, housing, and utility services. By contrast, households in the top quintiles are able to use a much larger share of their expenditures for goods and services that are neither food items nor expenditures for housing and utilities. For very poor households, non-food expenditures tend to shrink to expenditures for housing and utilities only. In extreme cases, the poorest of the poor may have to reduce their expenditures for food in order to be able to pay charges for housing and utilities. Alternatively, they may fail to pay for housing and utilities and become indebted to house owners, utility companies or municipal housing maintenance organizations.

Although DH tariffs have stabilized in those countries which have passed on the price increases of energy inputs and have adopted policies of full cost recovery, DH tariffs still represent a burden on household budgets and an even larger burden on household budgets of the poor. In other countries, such as Russia, where cost recovery will only be gradually introduced over the next 5 years, the burden of DH tariffs on household budgets will continue to grow. Thus, improving the efficiency of DH and hot water services in order to lower the costs of supply and thus improve their affordability is a major political issue and one of the highest priorities of local governments, the main owners of DH systems.

Impacts of DH on the Poor

The high cost of imported fuels for DH systems has led to a situation in many countries whereby heat and hot water services are being rationed, due to the poor affordability and non-payments problems which in turn do not allow utilities to purchase the necessary fuel for heat production. In countries with DH cities where the ambient temperature drops to -10°C or more, heating is a basic necessity of life as is food. The poor households suffer disproportionately from deficient heating and hot water service, as they have practically no margin for coping with the insufficient provision of these essential services. Higher income households are able to supplement DH with individual space heaters and by making improvements in windows and insulation, which the poor cannot afford. As a result, the poor have generally higher sickness rates and lower productivity.

To protect poor and vulnerable households against hardships in cold winters, a reliable provision of DH and hot water at satisfactory levels of quality is needed. Reducing heating costs for households through efficient investments is an important factor in ensuring adequate access, especially by the poor, to this basic energy service, by improving its affordability and thus the ability of DH utilities to secure necessary fuel supplies for sufficient service levels. Introducing modern DH equipment which allows buildings to measure and regulate the quantity of heat in-take also allows households to consume levels of heat and hot water better matched to the their abilities and willingness to pay. This is particularly important in Eastern and Central European countries where the large drop in real incomes raises questions about whether households need full levels of heating and hot water services. The current limited affordability of DH in

---

[48] Defined as the 20-25% of households with the lowest per capita expenditures.
[49] ESMAP, 1998.

some countries, such as Armenia, for example, has led to a policy of reverting back to traditional fuels for heat and abandoning DH. Investments aimed at improving the technical performance of DH can considerably improve the reliability of the heating service as well as substantially reduce the costs of supply.

Furthermore, the effectiveness and efficiency of social assistance programs need to be improved to reduce leakage of funds to non-eligible households and improve coverage to support low-income households in paying charges for heat and hot water services. Non-targeted privileges (discounts) for particular categories of households (police, military, etc.) need to be abolished and used for targeted social assistance. While a number of different subsidy mechanisms are in use in Eastern and Central Europe and the FSU, the mechanism that is most suitable for a particular area must consider the local circumstances and such information as the share of poor connected to DH, the possibilities that exist for proper estimation of actual household consumption and administrative costs, among others. It is expected that the requirements for subsidies to households are expected to continue over the next 5-10 years, until household incomes grow commensurately with the rise in service prices.

C. World Bank Experience in Financing District Heating Investments

Types of Projects

Financing district heating projects is a relatively new business activity for the World Bank Group. The Poland Heat Supply Restructuring and Conservation Project was the first such project approved for financing in 1991. It was designed to support the Government's Economic Transformation Program initiated in 1990. The energy sector, and DH in particular, was characterized by its high energy intensity, institutional inefficiencies and operational wastage, dependence on the state budget for investment and price subsidies, and the absence of market-related price signals. In addition, the energy sector was the country's major source of air pollution resulting primarily from burning coal for power generation and heating. The World Bank financed DH investments in four cities of Gdansk, Gdynia, Krakow and Warsaw in this first project and in Katowice in a later project in 1994 to address the technical and institutional inefficiencies and environmental problems.

This project was quickly followed by projects or project components in China, Estonia, Latvia, Bosnia, Krygyz Republic, Slovenia and Ukraine. In China, interest to introduce DH systems originated in order to reduce air pollution in very densely populated urban areas. The first request to the World Bank was to finance construction of a DH system in Beijing as a component of a larger Beijing Environment Project. A second project, the Shandong Environment Project, was initiated in Shandong Province in the cities of Weihai and Yantai to construct greenfield DH systems based on coal-fired CHP production. The projects were designed to eliminate the use of household stoves and a large number of small coal-fired HOBs which were contributing to a very high level of air pollution leading to, inter alia, respiratory illnesses in these cities. A number of similar projects are also being supported in China with financing from the Asian Development Bank and private commercial sources.

The request for World Bank assistance in Estonia in 1992 was a result of the Government's desire to improve the security of its heat supply systems after independence. Estonia's DH systems relied primarily on imported fuels, mainly gas and heavy fuel oil from Russia and coal from Poland. After independence, fuel supplies, mainly gas and heavy fuel oil, were intermittently disrupted. Estonia therefore wanted to utilize domestic fuels, wood products and peat, which are abundantly available all over the country, for heat production through an investment program of boiler conversions and replacements along with investments to improve overall system efficiency in the 3 largest cities of Tallinn, Tartu and Parnu as well as in small municipalities in the countryside. The project was co-financed by the European Investment

Bank and the Swedish Government and supported by donor organizations from Finland, Sweden and Denmark. The boiler program was also supported by the EU, EBRD and the Swedish Government through NUTEK through complementary efforts.

The project in Kiev, Ukraine was in response to the lack of adequate heat, which was a severe bottleneck to growth in the expanding capital city. At the city center, the capacities of the main boiler plants, which were nearly 50 years old, had been reduced due to the lack of rehabilitation and replacement investments. As a result, heat supply to households was being rationed and development of new and renovated office space, hotels and other buildings was severely constrained. In addition, new residential development, housing 260,000 people, in the last growth pole of the city was without any own source of heat. As an emergency solution, a pipeline had been constructed from the already overloaded CHP plant in the city center to the area, whereby households were being served a bare minimum level of heat (around +10°C). Kiev was facing a crisis situation with regard to heating and hot water services.

The project in Jelgava, Latvia was initiated to address the poor technical condition, high level of inefficiency and the short remaining lifetime of the DH system. The Bosnia project was an emergency project for reconstruction of essential infrastructure, including DH, after the war. The project in the Kyrgyz Republic was designed to address the key inefficiencies in the heat and power system in the capital city of Bishkek where the Bank played a coordinating role in the implementation of a project co-financed by the Asian Development Bank and the Danish Government. The Slovenia project was driven by environmental considerations and included a component to convert the heating systems to cleaner fuels. Other complementary projects or project components were also undertaken in Lithuania, Russia and Czech Republic.

Most of the projects being supported by the World Bank involve the rehabilitation and modernization of existing DH systems. The exceptions are the China Shandong Environment Project, which financed greenfield DH systems, and the Ukraine Kiev District Heating Improvement Project, which financed completion of a DH network and production capacity in the last growth pole of the capital city. In these cases, DH has been shown to be the least-cost long-run heating option for the concerned areas. All the projects, both rehabilitation and greenfield, generally included packages of institutional support and pricing reforms to provide the proper incentives to encourage energy conservation.

Complementary projects to reduce energy wastage in buildings, also known as demand-side management projects, are being undertaken in Lithuania, Russia and Ukraine and are under preparation in other countries such as Poland.

A number of early projects have been successful in mobilizing grants from the GEF to complement other sources of financing for improving the efficiency and environmental conditions of DH systems in Poland, Lithuania and the Czech Republic. These projects have focused on utilizing cleaner fuels, such as gas and geothermal sources, or waste heat to maximize the level of carbon savings in heat and electricity production, important for mitigating climate change. In addition, these projects include other investments to improve overall DH system efficiency. Further projects with GEF components are under preparation in Poland, Hungary, the Slovak Republic and Slovenia.

Today, further projects are under preparation or are nearing the stage of implementation in a number of other countries, including Bulgaria, Croatia, Russia, Latvia and Lithuania, for example. One project, which is under preparation in Sevastopol, Ukraine, would support investments in a decentralized heating system based on gas-fired mini-boilers, since DH was shown not to be the least-cost heating alternative. A summary of the World Bank-supported projects, both under implementation and preparation, is presented in Annex 8.

World Bank Instruments and Lending Volumes

To-date, the key instrument of the World Bank Group for assisting countries to improve their heating systems and the energy efficiency of buildings has been the World Bank's sector investment loan (SIL). SILs have been shown to be highly successful in establishing policy changes, institutional reforms and technical improvements in the early projects completed, which are now being emulated elsewhere.

In a few cases in Poland and Estonia, credit lines for DH improvements were experimented with, but with mixed results. The credit line approach was not successful in Poland for a number of reasons, including: (a) the poor financial condition of DH enterprises which made them not creditworthy borrowers, (b) project selection procedures and procurement rules were cumbersome, and (c) many enterprises were undergoing ownership changes during which they were prevented from borrowing by their owners. On the other hand, in Estonia the credit line was a successful vehicle for channeling loans to small municipalities in the countryside for DH boiler investments, as it was supported by a central mechanism established for assisting in the development of feasibility studies and to properly screen the projects. In addition, procurement requirements were simplified given the small size of the investments.

Structural adjustment loans (SALs) have not been utilized to-date to address deficiencies in DH and energy efficiency and are not particularly appropriate for this purpose. This is due to the fact that SALs are typically utilized to support nationwide reforms under the responsibility of the central government, whereas DH and energy efficiency are more a matter of local governments and agencies.

Since 1991, the World Bank has been working together with GEF to mobilize additional funding to assist in the protection of the global environment and to promote environmentally sound and sustainable economic development. GEF grants are used to meet the incremental costs of measures to achieve agreed global environmental benefits in the areas of climate change and ozone layer depletion, among others. As of June 1999, 227 projects with a value of US$ 884 million in the area of climate change and another 17 projects with a value of US$ 184 million in the area of ozone depletion have been approved. These funds have helped to mobilize additional resources from other public and private sources.

More recently, the World Bank has established a Prototype Carbon Fund (PCF), with the objective of mitigating climate change. The PCF will pilot the production of emission reductions (ERs) within the framework of Joint Implementation (JI) and the Clean Development Mechanism (CDM). The PCF will use the contributions made by companies and governments to purchase ERs through support to specific projects designed to produce ERs credibly and additional to those that would occur under business-as-usual financing.

Altogether, the World Bank has up-to-now provided loans in excess of $1.3 billion towards the cost of the early DH and energy efficiency projects, which catalyzed another $1.7 billion of funds for the investments and institutional support.

IFC's relevant experience in financing DH investments has been nil while its experience in financing energy efficiency investments has been concentrated mainly on development of private energy service companies (ESCOs), financing projects amounting to about US$ 127 million in a number of countries including Hungary and Poland. IFC also works together with the GEF, and occasionally with other multinational and bilateral sources, to access concessional funding in cases, for strategic and/or developmental reasons, that justify the use of below-market rate financing for the penetration of renewable energy and other energy efficiency activities that have a high likelihood will continue in the future on a commercial basis.

Project Results

The Poland projects are now completed and their results have surpassed expectations in terms of financial, economic and environmental benefits obtained and pricing, policy and institutional reforms achieved. A summary of the results of the Poland projects is included in Annex 9. Similar results have also been obtained in the Estonia project. Some of the interesting outcomes which have already been achieved in these early projects include the following:

- In Poland, at the end of the implementation period of the project, the price for heating one square meter of floor area decreased for DH consumers by 55% in real terms as a result of the efficiency gains from the project investments.

- In Tallinn, Estonia, completion of a pipeline interconnection between two isolated DH networks allowed for a greater use of CHP heat, a higher level of CHP capacity utilization and a lowering of the bulk heat tariff from the CHP plant by about 15% for the incremental heat purchases.

- The Estonia program of converting HOBs to use renewable fuels (peat, wood chips and wood waste) instead of imported fossil fuels in two large cities and 36 small municipalities in the countryside, together with other investments supported by other funding sources, has helped to increase the share of renewable energy in Estonia's primary energy balance from 3.5% in 1993 to 11% in 1998.

- In Poland, investment subsidies were eliminated, and household subsidies phased out gradually from a nationwide average of 78% of the heating bill in 1991 to zero in 1998.

- In Kiev, Ukraine, heat tariffs for consumers were reduced by about 20% during project preparation as a result of undertaking a heat tariff study that demonstrated how costs could be reallocated in the CHP plants to allow for a greater sharing of the benefits of the co-generation process with heat consumers and for maintaining competitiveness of the DH service.

- In Jelgava, Latvia, installation of automatic consumer substations and heat meters allowed consumers to save about 25% of the heat energy previously consumed.

- In Krakow, Poland, air quality was improved through reduction in gaseous and dust emissions as a result of the elimination of polluting HOBs and the coal-to-gas boiler conversion program, which allowed the Krakow heating company to be officially eliminated from the list of heavy air polluters.

- Rehabilitation of the Estonia DH systems in the 3 largest cities has led to an increased interest by private investors to acquire assets in the DH systems, and parts of 2 of these systems have been privatized through a competitive process attracting foreign strategic investors which have a keen interest in privatization of the remaining parts of these system.

- In Riga, Latvia, the prospect of implementing a DH rehabilitation project, which would increase the electricity (and heat) production at the existing CHP plants to a level that would allow Latvia a greater degree of self-sufficiency in electricity production, has resulted in lower electricity import prices by about 15%, even though project implementation has not yet commenced.

- In both Poland and Estonia, the rehabilitation projects have contributed to the development of local manufacturing industries for energy equipment (e.g., DH pipes, substations and heat meters) and installation companies, with these supply industries competing for and winning a significant share of the markets in both Western and Eastern Europe.

74

- The switch to biomass as a fuel in heat production in Estonia has led to the creation of employment opportunities related to the harvesting, transport and disposal of these fuels in rural areas, with an estimated 200 permanent jobs alone in the area of Tartu.

- For the first Poland project, the overall ex-post economic rate of return was 48% without environmental benefits and 88% with environmental benefits, as compared to an estimated ex-ante 30% economic rate of return.

These results have confirmed that DH rehabilitation and improvement projects and other types of energy efficiency investments can be successful even in risky countries. The human capital and technical skills of energy utility managers and staff and policymakers are high in Eastern and Central European and FSU countries, and these factors have worked to greatly reduce implementation risks. Furthermore, these types of projects are not linked to a large nationwide reform program, such as in power, and therefore have higher probabilities for success. DH and energy efficiency together represent an excellent area where the World Bank can diversity its portfolio and reduce the overall risks.

D. Rational for World Bank Involvement in District Heating or Other Heating Systems

Huge Investment Requirements

As mentioned previously, least-cost rehabilitation investments in DH systems, or in alternative decentralized heating systems where DH is not the least-cost option, can bring about substantial benefits resulting from reductions in fuel usage and environmental emissions, and the requirements for funds for these investments are huge. An estimate of the order of magnitude of investment needs has been roughly calculated, based on the requirements for investments which were needed to bring about at least a 20% energy savings in the Bank's recently completed projects in four Polish cities. For these projects, a regression analysis was carried out utilizing the project costs and the heat sales of the individual systems to compute a formula which could be applied to other countries where data on heat sales is available.

For 11 countries in Eastern and Central Europe and the FSU where such data is available, including Russia, Ukraine, Romania, Poland, Czech Republic, Hungary Lithuania, Estonia, Bulgaria, Croatia and Slovenia, the total annual heat sales is about 3,700 PJ and investment requirements, excluding those already undertaken, are estimated to be about US$ 25 billion over a 5-7 year period. When the remaining 15 Eastern and Central European and FSU countries are considered, the total investment requirements would be considerably higher. It has been further calculated for these 11 countries that a 20% savings in fuel requirements would lead to an annual reduction of 24 million tons of carbon emissions and an annual savings of about 8 billion cubic meters of gas, 7 million tons of heavy fuel oil and 23 million tons of coal, assuming a fuel mix of gas, heavy fuel oil and coal at 25%, 25% and 50%.

Limited Availability of Financing for DH Investments

Since DH companies in Eastern and Central Europe and the FSU have not been able to build up adequate reserves for future investments through their tariff policies, funds for investment requirements must be raised from external sources, such as loans or private equity. However, the possibility to raise funds from these sources at terms suitable for infrastructure investments is virtually non-existent in Eastern and Central Europe and the FSU. The local commercial banking sector in many countries is still underdeveloped, but even where local loans for investment purposes are available, they are usually only available at high interest rates and with short maturities - terms which do not match the requirements of typical DH projects where the investments have long lifetimes. In the secondary cities where DH systems are smaller, less efficient and typically do not have access to heat from CHP plants, the availability of commercial bank financing is even more limited.

The experience to-date with raising equity from private investors is also very limited. So far, interest from private investors has been concentrated on CHP plants, which are typically separate enterprises from the DH network utilities, as in Hungary and Czech Republic, or on electricity enterprises which also include the DH business, as in Kazakhstan.

Sustainable private participation in infrastructure services requires certain conditions which are generally lacking in Eastern and Central European and FSU countries. At the macro level, the factors needed for private investment include political stability, currency stability, foreign investment laws, convertibility of currencies, transparent and independent regulatory framework, reasonable tax regimes, commitment of the government to non-interference and sectoral reform, access to consumers for revenues, potential for "exit" from the investment, among others, but many countries are not yet able to offer these to potential investors. At the sector level, the potential to recover costs and to exert leverage over customers to encourage payment discipline remains difficult. Given that DH tariffs remain even today the highest or next highest expenditure in a household budget limits the willingness and ability of customers to pay for the DH service. Even if customers are willing and able to pay for this basic service, the inability of most DH enterprises to disconnect customers because of the lack of metering, the use of intermediaries in billing and the lack of a direct relationship with the customers leaves DH enterprises little leverage over customers to encourage payment.

In this sense, the DH sector differs significantly from the electricity sector in Eastern and Central European and FSU countries, which typically already have installed electricity meters and have the ability to disconnect consumers which do not pay. This may help to explain why private investors who are willing to enter countries in Eastern and Central Europe and the FSU in the electricity sector may find the DH sector unattractive. More specifically, the interest of private investors in CHP plants may also be more related to having off-take agreements and government guarantees than to the cost recovery potential of the heat and electricity retail businesses. Since DH is not common throughout the world as compared to other municipal services such as gas, water and transport, the absence of a project demonstrating the privatization of DH services and the lack of industry specialists and financiers with privatization experience in the sector may also be reasons why private sector participation in DH is not a common initiative. The United Kingdom, New Zealand and Latin America are the primary examples for the initiation of private infrastructure programs but none have DH except the United Kingdom but only to a minor extent.

Estonia and Poland offer a few examples with private participation in the provision of DH services. In Estonia, the private sector has recently acquired and is now actively competing for ownership of DH utilities in Tartu and Parnu, but this has occurred only after the World Bank-financed projects have been completed, heat meters have been installed, and the DH utilities are operating on a commercial basis. In Poland, private participation involved acquisition of shares by strategic investors or through some performance contracting principles ("contract d'affermage"). However, this experience is relatively recent, and the technical benefits and operational efficiency gains have yet to be clearly demonstrated.

In the FSU and Eastern and Central European countries, MDB financing is therefore likely to remain as the main financing instrument over the near term as a consequence of their still relatively less developed economies, poorly developed capital markets, lack of access to long-term financing and limited opportunities for attracting private investment.

A Special Role for the World Bank

Development of Least-Cost, Long-Run Heating Strategies. The World Bank can play an important role in assisting Eastern and Central European countries, both at the country-level and at the city-level, to develop heating strategies, which take into account the factors necessary to determine the least-cost, long-run alternative(s) most appropriate for those areas. DH systems were the preferred and even mandatory approach to heating in Eastern Europe, but DH has not always been the least-cost solution, as has been shown in Orenberg, Russia and Sevastopol, Ukraine. Master plan studies or country-wide strategy studies can be designed to look beyond current energy price levels and short-term distortions to determine the optimum long-run option for heating.

Where DH is the least-cost alternative, countries and cities can benefit from assistance in identifying investments which allow for the optimization of CHP plants, typically outside the purview of DH enterprises and their municipal owners and which have, not uncommonly, been excluded from the operation and investment planning of those enterprises. Within DH systems, benefits can also be achieved by analyzing the various areas of a city to determine where DH may not make sense and should be closed down with decentralized options installed in its place. Furthermore, in areas where DH is not the optimum alternative, benefits can be achieved from assistance in identifying a more appropriate decentralized option, based on the fuel options available and other factors which influence the performance of heating systems and their viability. The World Bank has now developed considerable expertise to be particularly effective to help guide such strategy studies in these areas.

Poverty Alleviation. The World Bank, through its heating investment projects, can also directly assist poor households connected to DH or utilizing other heating options in a number of ways. First, by participating in the design of projects which would reduce the cost and improve the quality of energy supplied, and thereby lower heat tariffs and improve affordability of heating services, the Bank is helping to improve and better ensure access to heating and hot water by the poor. The first Poland DH project, where DH tariffs were reduced by 55% in real terms over the project period as a result of the efficiency gains, and the Kiev DH project, where DH tariffs were reduced by 20% during project preparation by reallocating costs in the CHP process, are two very positive examples of how affordability and access to DH can be improved.

Secondly, through investment projects which improve heat service reliability and affordability, household expenditures can be lowered for supplemental heating, and resources by all households, including poor households, can be increased for other energy efficiency investments, such as weather-stripping of windows and doors, inside apartments and insulation of roofs inside buildings which further help to reduce heating bills.

Thirdly, by working together with central and local governments to improve the targeting of energy subsidies, the Bank is helping to better ensure that social assistance reaches the poor and vulnerable groups.

Fourthly, by promoting energy efficient and less polluting end-use technologies, the Bank is working to improve health conditions affecting both the poor and other households.

Given that support for the poor in Eastern and Central Europe is estimated to require an increase in assistance over the next 5-10 years, a research paper on "Heating Options for the Poor" is planned to be undertaken by the World Bank during 2001 which will further analyze how poor households are affected by heating services and identify the best measures for assistance.

Supporting Macro-Fiscal Stabilization. World Bank-supported heating projects in Eastern and Central Europe can lead to significant macro and fiscal impacts. Fuel savings through efficiency improvements have now been demonstrated to typically amount to more than 25-30% in the early projects, and these savings translate into direct improvements in the balance of payments for fuel importing countries.

Heating-related expenditures are a high burden for many municipal budgets, and increasing efficiency and reducing costs of heat supply would go a long way in reducing the burden on municipalities. These types of projects have been shown to provide significant relief on municipal budgets, by providing financing for urgently-needed capital investments and by reducing the level of operational subsides needed by heating utilities as operational costs are reduced through efficiency improvements. Reducing the costs of heat supply also reduces the level of subsidies needed by low-income and vulnerable households, thereby further reducing the level of subsidies to be provided by local budgets as well as by the central government budget.

Promoting Environmental Sustainability. The World Bank can play a leading role in helping countries in Eastern and Central Europe to protect the environment through investment projects in DH, or in decentralized heating systems, where the scope for reducing harmful emissions is high. As mentioned above, the improvements in efficiency in the early World Bank-supported DH and energy efficiency projects have resulted in significant reductions in fuel use, leading in turn to reductions of emissions, particularly CO_2 and other greenhouse gases, important in mitigating climate change.
The early projects, especially the Poland and Estonia DH projects, have also helped to protect the environment by promoting the switch away from coal and heavy fuel oil, to cleaner fuels, such as gas or wood products, in heat production and by including measures to improve pollution controls.

The Estonia DH project and other projects often supported by the GEF have also been important in assisting to remove the barriers to renewables and energy efficiency investments, further helping to protect the environment. Through such investments, the World Bank is further assisting countries in Eastern and Central Europe, many of which are in the process of EU Accession, to move in line with EU policies promoting greater energy efficiency and environmental protection.

Furthermore, through its recently-established Prototype Carbon Fund, the World Bank will pilot the production of emissions reductions within the framework of Joint Implementation and the Clean Development Mechanism. Substantial opportunities for trading carbon credits exist in the area of DH and other types of heating investment projects in Eastern and Central European countries.

Improving Regulation of Heat Pricing. By applying its global experience in the heating sector, the World Bank is also helping countries to develop appropriate pricing policies and to establish a clear and independent regulatory framework to improve energy regulation. This is especially important in the areas of bulk heat pricing from CHP plants, where the benefits of the co-generation process are generally not shared with DH, and gas pricing, where lack of differentiation of prices among consumer groups, with small consumers typically paying the same as large consumers, creates unfair competition to DH. The World Bank has been proactive with its clients and the regulators in Eastern and Central Europe in working towards full cost recovery of heating services, the elimination of cross-subsidies in heat tariff structures and the introduction of two-tier heat tariffs to promote greater transparency in heat pricing. Clear and transparent regulation of heat tariffs, in turn, helps to reduce an important barrier to private sector involvement.

Demonstration Impacts with Spillover Effects. While the early World Bank projects in DH have been focused on particular localities, these early projects have had important demonstration effects in their respective countries. The recently-completed DH rehabilitation project in Jelgava, Latvia, for example, has led to interest by other cities to rehabilitate their heating systems, and major investments are planned or ongoing in other areas today with various sources of financing. The publicity about the Kiev DH project has also led to interest by a multitude of localities to improve the efficiency of their heating systems, and many systems are under review to determine the feasibility of investments. This interest is expected to lead to a number of projects being replicated in the near term.

Catalyzing Additional Funding for Heating Investments. The World Bank-supported interventions in DH and energy efficiency are helping to mobilize co-financing from other international financial institutions and donor organizations. By taking the lead in helping to set the overall policy framework and build the institutions in the sector, the World Bank has been able to attract substantial additional financing from such sources as EBRD, EIB and NIB as well as from bilateral aid and export credit agencies. In addition, where the World Bank is involved providing comfort in project design, other donors with grant funds, such as GEF or bilateral donors, are also participating. In this way, the share of World Bank financing of investment projects can be reduced as other organizations take a larger role in financing.

Supporting Private Sector Development. World Bank-supported investment projects in DH and energy efficiency are starting to demonstrate the efficiency gains and institutional and financial improvements which can be achieved by DH enterprises, and other heating agencies, in countries in transition. The World Bank can thus be seen as paving the way for greater commercial financing and private sector investments, as DH and other heating utilities are commercialized, which reduces the risks and makes the heating sector more attractive for private financing. The results of the recently-completed projects in Poland and Estonia show that the private sector is willing to invest in DH, once the economies develop to a stage where adequate legal and regulatory frameworks are in place and the energy utilities have been commercialized and corporatized. The upcoming Riga, Latvia DH project has also demonstrated that commercial banks are willing to participate in financing investments at suitable terms when the World Bank is helping to establish the policies in the sector and the DH enterprise is operating on a commercial basis. These results are encouraging for fostering further private sector involvement in the future.

Conclusion

As this report has shown, the World Bank is well positioned to respond to the opportunity and challenges posed in the energy sectors in Eastern and Central Europe and the FSU today in a number of ways. The World Bank's involvement in a variety of early DH and energy efficiency projects has clearly demonstrated the Bank's special role in these fields. These early interventions have taken place at a time when these countries were facing severe resource constraints and lacked access to private capital at appropriate terms. Based on the early results, the World Bank has shown that it can play an important role during the transition period to market economies by providing capital to support the much-needed policy and institutional changes and heating sector investments until the sector is able to attract a sufficient volume of capital from other sources.

Summary of the Extent and Characteristics of Western European District Heating Systems

Iceland

The DH system in Iceland is unique. DH was introduced in the 1920's and during the period of 1950 to 1980. DH became increasingly popular and today more than 85% of the homes are connected to the DH networks. Iceland has large amounts of geothermal energy and bases almost all (96%) heating upon this source; the remaining 4% is produced in hydro-powered electrical boilers.

Denmark

Danish DH dates back to the turn of the century. It was initially proposed to reduce the cost of supplying space heating. Prior to 1950, almost all DH was supplied with heat from municipal waste incineration and electricity production. In the period of 1955-73, more than 200 new independent DH networks were established all over Denmark. During the period, oil was used to a large extent for DH. After the sharp oil price increases in 1973-74 and 1979-80, a large proportion of fuel oil-based DH companies opted for the conversion of their plants for use of other fuels.[50]

Originally, only buildings close to power plants were supplied with DH, but the DH networks were gradually extended to take in areas at a greater distance. Factors of economy and security of supply were the principal considerations that spurred the strong expansion in using surplus heat from power stations. In recent years, however, improving the environment has gained growing importance.

The market share of DH of the whole building stock is 49% followed by light oil heating at 22% and natural gas heating at 18%.

Today, there are 58 public DH companies operating in the larger cities and about 340 private consumer-owned companies operating in Denmark. About 250 of them produce electricity as well. In large towns, power and DH are often united in a public company. Currently, approximately 90% of Danish DH production is based on either biomass fuels or coal-fired CHP. The share of CHP of the total DH production is 61%.

Finland

In Finland, the first installations were commissioned in Helsinki in 1940. Wider introduction began in the 1950's, but the speed of introducing new schemes was really triggered by the 1973 energy crises and the years of 1973-85 were time of rapid expansion. Today, 50% of all buildings and 55% of homes are connected to DH; the share is even higher in some cities, like Helsinki, where the share of DH is 90%.
In 1995, over 75% of DH was produced in CHP plants; in 1997, the share had increased to 79%. The main fuels are coal (36%), natural gas (29%), peat (22%) and oil (6%). Currently, the share of natural gas is increasing whereas the share of coal is decreasing.

The main competitors of DH in the existing building stock are light oil heating with a share of 22% and decreasing, electric heating with a share of 15% and wood with a share of 11%, mainly in rural areas. In

---

[50] Ministry of Energy and Danish Energy Agency, 1993.

new buildings, the share of DH is 49%, light oil 13%, electricity 22% and others 16%. The price of electricity is two times as high as the price of DH, but the lower investments needed for electric heating make it competitive, especially in smaller buildings.

Sweden

In Sweden, the first DH network was introduced in 1948 in Karlstad. The relative prices of fuels have been favorable for DH development. Heavy fuel oil used in power plants was much less expensive than light fuel oil used in individual heating systems. Therefore, it was more profitable and competitive to invest in DH. Later, heavy fuel oil was substituted by fuels with less taxation, other generating methods and heat purchases from industry.

Today, the market share of DH of the whole building stock is 38% (36% in 1995), followed by light oil heating with a share of 27%, electric heating with a share of 26% and solid fuels including wood with a share of 10%. In new apartments, the shares are significantly different: DH 70%, light oil heating 10%, electric heating 18% and wood etc. 2%. The wood pellet boiler is a growing competitor in single-family houses.

Biomass is the most important fuel for DH and CHP production with a share of 28% (20% in 1995) and other domestic fuels, such as peat and refuse, comprise about 16% of the DH supply. 7% of DH supply is derived from waste heat. 28% of DH is produced in CHP plants.

Austria

The first attempts to introduce DH in Austria were by CHP plants before World War II, but they were restricted to specific industrial systems. In 1949 the first local CHP plant was put into operation in Klagenfurt, which was followed by a few more during 1955. Despite the low prices of DH's competitors, DH could steadily increase its importance due to the objectives of energy and environmental policy.

Almost 400,000 flats, representing 12% of the whole stock, were supplied with DH in 1997. Market share has been growing and by 1998 it had increased to 12%. Currently, DH focuses on the cities of Vienna, Graz, Linz, Salzburg, Klagenfurt, St. Pölten and Wels.

Germany

The share of district heated homes in the old federal states was 10% in 1995 and 12% in 1996; the share was 22% in new federal states in 1996. The share of different consumer groups of the total consumption is: 46% households, 36% public installations, business and commerce, and 18% industrial customers. Currently, DH focuses on the cities of Berlin, Dresden, Essen, Hamburg, Mannheim and Münich. DH is in direct competition with natural gas heating with a share of 37% and fuel oil heating with a share of 35%. In the new building stock, the shares are natural gas 58%, fuel oil 30% and DH 7%.

64% of DH was produced in CHP plants in 1996. In 1996, the main fuels were coal (48%) and natural gas (41%) followed by oil (5.5%) and refuse (4.5%). In CHP plants, the fuels are coal (68%), natural gas (20%), fuel oil (7%), and others (5%). The high share of coal results from German energy policy concerning mineral coal and the increased lignite input in the eastern part of the country.

Today, Germany publishes statistics which are combined for the eastern and western parts of the country, but this doesn't allow the specific differences in the DH systems of the two areas to be shown. The reconstruction process which has been initiated in the eastern part of the country, after several years, is expected to bring about comparable results with the western part of the country.

France

After the first installation in Paris in 1928, the use of DH increased to a larger extent only during the 1960's; some 150 DH systems were in operation in the early 1970's.[51] Today, the most important DH systems in France are located in Bordeaux, Chambery, Grenoble, Metz, Paris and Toulouse. DH has an overall 3.5% market share. The competition is intensive between electricity, natural gas and DH.

Two-thirds of DH production is based on coal, fuel oil and natural gas. About 17% (14% in 1995) of DH production takes place in CHPs. One-third is based on geothermal energy, refuse incineration and waste from industry. The introduction of new schemes is often based on establishment of refuse incineration plants.

The Netherlands

Thirty years ago, the fuels for space heating were coal and oil. Now, all the coal mines are closed and natural gas has become the most important fuel ever since large indigenous resources were found. In the past 25 years, about 96% of all houses in the country have been connected to the natural gas network.

In the Netherlands, the number of district heated homes in 1997 was about 200,000, representing 3% of the whole stock. The share in the services sector (offices, schools, hospitals, universities etc.) differs from the housing sector. In the major city, the Hague, 80% of the offices, including government buildings and the Parliament, are heated by DH. The current policy of DH companies is to connect larger entities, such as office buildings, hotels and universities.

In 1997, about 90% of the DH was produced in CHP plants. The main fuel was natural gas at 99% (97% in 1995).

Norway

Heating based on occasional power with light fuel oil as backup has traditionally been a serious competitor to DH. The reason is the low price for occasional power due to surplus production from hydropower stations. New buildings are usually equipped with direct electric heating since this demands a lower initial investment than a water-based heating system.

Today, DH accounts for 3% of the heating of buildings. The share of electric heating is 76%, fuel oil 12% and wood 9%. The share of electric heating has grown from 67% in 1995. The main sources for DH production are waste 49%, oil 20% and electricity 11%.

The most important fuel for DH production is refuse, followed by electricity, industrial waste heat, wood chips, light fuel oil, gas and heat pumps.

Italy

The first DH scheme was introduced in Milan in 1965.[52] In Italy, the market share of DH is 1.5%. The use of DH is limited in absolute terms but the increase has been dramatic; the consumption grew 4-fold during the ten years from 1986 to 1996. The number of towns with DH is 27; the largest schemes are in Brescia, Torino, Verona, Reggio Emilia, S. Donato M. and Mantova. The main competitors are natural gas (77%) and oil (21%). Natural gas is the most important fuel in DH and CHP production with a share of 70%. About 75% of DH is produced in CHP plants.

---

[51] Huovilainen and Koskelainen 1982.
[52] Huovilainen and Koskelainen, 1982.

United Kingdom

The first DH schemes were installed in Glasgow, Manchester, Dundee and Chesterfield in the early 1900's. DH development escalated after World War II, and the busiest period was during the 1960's and 1970's when over 500 small-scale schemes were installed. The popularity of DH began to wane following the oil crisis of 1973, because many of the schemes were oil-fired. This was combined with the decline in local authority housing construction in the late 1970's and 1980's. Consequently, the expectations for rapid expansion of DH never materialized. In the United Kingdom, DH is usually called community heating. The new name was launched partly because, after the oil crisis, DH had a negative image.

The distribution among different DH consumer sectors is: health care and hospitals (80%), industrial buildings (41%) and housing (10%).

Today, there are some 250,000 homes, representing 1% of the United Kingdom housing stock, connected to about 800 DH schemes. Therefore, the average size of one scheme is very small. Most operating schemes are in local authority housing which, together with the housing association sector, has the most potential for further community heating. New schemes include London (the City, Waltham Forest, Southeast London, Camden), Southampton, Glasgow, Edinburgh, Nottingham, Sheffield, Manchester and Birmingham.[53]

Currently, under 5% of the DH schemes have CHP. During the period 1988-95, the installed CHP capacity has almost doubled, representing an average growth rate of 9% per annum. In 1996, the growth rate stayed at 2%. The installed capacity was 3,562 MW of electricity in 1,336 sites in 1996. Consequently, the average size of installations is very small.[54]

Switzerland

The first DH schemes, introduced in Lausanne in 1934, in Basel in 1943 and in Bern in 1954, used heat from waste incineration plants. The next installation was in Geneva, motivated by environmental considerations.[55] DH has grown only modestly in Switzerland in recent years. The market share of DH is only 2%. The share of CHP for DH production is 32%. There is very little incentive to increase the share of DH and use of CHP due to the low prevailing electricity prices. The main fuels for DH production are gas (32%) and refuse (49%).

Greece

The two first DH systems were installed in Kozani and Ptolemais in the beginning of the 1990's. After the good operating experience and recorded reduction in emissions, two more systems, Florina and Megalopolis, will be installed in the near future. Currently, the dominant heating form is individual heating by diesel oil.

---

[53] Green, 1995.
[54] GSS, 1997.
[55] Huovilainen and Koskelainen, 1982.

Key New Merchant District Energy Systems in the United States

Trigen

An independent energy supplier, Trigen develops, owns, and operates commercial and industrial energy systems in North America. The Company serves about 1,500 customers with energy produced at 31 plants in 23 locations. The first Trigen project was the Trenton, New Jersey "trigeneration" project, which heats and cools 38 buildings in the heart of downtown Trenton. The project uses two diesel generators fueled primarily by natural gas to produce 12 MW of electric generation, 190 GJ of thermal heat and 8,500 tons (1 ton of cooling = 12.72 MJ) of chiller capacity. It sells the electricity to the local utility. The project had originally experienced financial problems which were solved by the addition of a DC system.

Since 1995, the partnership of Trigen with Peoples Gas Company has been supplying DE to the McCormick Place Convention Center in Chicago, Illinois. In order to reduce the peak electric demand charges, the DE company installed a 39,000 m^3 stratified chilled water storage facility. A pair of 600-ton gas-fired absorption chillers feeds the thermal storage equipment. The cold water storage facility furnishes 135,000 ton-hours of cooling. The project created a competitive environment in the city of Chicago causing the local electric utility company to enter the DE market in order to retain their electric customers.

In 1997, Trigen in partnership with PECO Energy and NRG Generating started the operation of the 150 MW co-generation plant (Grays Ferry Co-generation Project, GFCP) in Philadelphia, Pennsylvania. The co-generated steam is supplied to the downtown district steam system owned by Trigen to 390 customers. It took seven years to develop the project. Shortly thereafter, PECO Energy notified GFCP that, because of the deregulation of the electricity market, the utility would only pay market rates for the electricity delivery by GFCP. The suit, filed by NRG and Trigen Corp., asked for an injunction to reinstate the power purchase agreement (PPA). The court ruled that PECO must honor the terms of the PPA and GFCP is to receive injunctive relief during the period of litigation to sustain the project.

Trigen Energy Baltimore (TEB), a wholly-owned subsidiary and the regional company for Trigen Energy Corporation, provides energy services to more than 350 buildings in Baltimore. The company owns the old district steam system, previously operated by the local electric utility. Trigen owns and operates four DE plants in Baltimore. TEB was recently selected as utility agent by H&S Properties Development Corporation to provide energy to occupants of the 18-acre site at the Inner Harbor East Development property in downtown Baltimore. 10,000 tons of cooling and 95 GJ per hour of hot water will be available to heat and cool the new customers. The value of the services is estimated at $150 million over 20 years.

Trigen-Cinergy Solutions, in partnership with District Energy St. Paul, has been selected by Northern State Power to produce via CHP generation, 25 MW of electrical energy fueled with local, urban wood waste. As part of the long-term PPA, the project will also provide additional hot water to District Energy St. Paul's heating and cooling system.

Trigen-Cinergy Solutions of Cincinnati has recently constructed a new cooling system to provide services at the new Paul Brown Stadium.

Trigen-Colorado acquired the Coors energy facility in Golden, Colorado in 1995. Trigen's plan includes expansion of the facility to provide electricity for additional Coors processes. The planned expansion put Coors in a position of strength during negotiations with their local utility, since they now had an alternative choice for energy supply. The resulting attractive new energy rate eliminated the need for expansion.

Trigen-Kansas City Energy Corporation in Downtown Kansas City, Missouri began early in 1998 the construction of a chilled water system to supply Government Buildings Downtown Kansas City. The total investment in the system is expected to reach $25 million by the year 2000. In 1990, Trigen acquired the downtown district steam system from the local electricity utility.

Trigen Energy and Stockholm Energi AB of Stockholm, Sweden entered into an agreement to extend Trigen's 11 urban steam systems with hot water distribution to the less dense areas of each host city.

In Champaign, Illinois, AC HUMKO, Illinois Subsidiary of Associated British Foods (ABF), will turn over ownership, operation and maintenance of its energy assets to Trigen-Cinergy so that Trigen may supply steam, process chilling, on-site co-generated electricity and compressed air to the manufacturing facility. Trigen-Cinergy will invest more than $11 million in upgrading energy equipment.

Trigen-Biopower, Inc, purchased Power Sources Inc. (PSI) of Charlotte, North Carolina. PSI produces energy from wood residues and other renewable biomass fuels. For five of its facilities, PSI receives fuel supplies from 120 waste wood suppliers. PSI saves landfill space by turning these waste products into energy.

Trigen's operations in Canada include Trigen Energy-PEI in Charlottetown, Prince Edward Island and Trigen-London (Ontario). These operations serve about 70 customers each. The generating equipment varies from wood waste-fired boilers and energy-from-waste operations to gas turbines.

In March 1998, Trigen Energy signed an agreement with leaders of the Navajo Tribal Utility Authority (NTUA), based in Fort Defiance, Arizona to provide 20 MW of electric power to NTUA and 50,000 kg/hour of process steam for a food processing facility. Trigen will serve NTUA from a CHP plant consisting of gas turbines, heat recovery steam generators and back pressure steam turbines.

NRG

A wholly-owned subsidiary of Northern States Power Company (NSP), NRG is one of the world's leading independent energy producers, specializing in the development, construction, operation, maintenance and ownership of co-generation and electric generating plants. Established in 1989, NRG is involved in independent power projects and thermal operations in the United States, Europe, the Pacific Rim and Latin America. NRG is the second largest DE provider in the United States. NRG owns, operates and manages DE systems in Minneapolis, San Diego, San Francisco and Pittsburgh (50/50% ownership with Thermal Ventures).

NRG recently purchased Energy Center Kladno (ECK), an energy complex that supplies 28 MW of electricity and 150 MWt of steam and hot water in the Czech Republic. The plan is to expand the facility with a 354 MW coal and gas-fired addition. Construction was scheduled for completion in 1999.

IPALCO

IPALCO's principal subsidiary is the Indianapolis Power & Light Company (IPL), which has been supplying electricity and steam in Indianapolis, Indiana for many years. IPL's district steam heating system began operation in 1889. Since 1989, IPALCO has expanded its unregulated DH/DC subsidiary, Mid-America Energy Resources (MAER). IPALCO has invested more than $60 million in DE capital expansion in the past three years and plans more than $75 million in additional capital investment over the next five years. Since 1991, MAER has owned and operated a DC system in downtown Indianapolis. The plant is supplying 33,000 tons of cooling.

Another subsidiary of IPALCO, Cleveland Thermal Energy Corp. (CTEC), operates a district steam system in downtown Cleveland, Ohio. IPALCO purchased this system, with 230 steam heating customers, in 1991. The steam generating capacity is 600,000 kg/hour at 1 MPa (unit of pressure; about 10 bar). Another subsidiary, Cleveland District Cooling Corp., was formed in 1992 and has an installed cooling capacity of 10,000 tons capable of growing to 25,000 tons.

Indianapolis Campus Energy (ICE) was formed to design, build, own and operate DE production facilities for large industrial, commercial and institutional customers. In 1996, ICE started the operation of a 15,000 ton cooling system for a new Eli Lily plant. The $21 million facility was built with provisions to double in size.

Unicom Thermal Technologies

UTT is a subsidiary of Unicom Corporation (and also the parent of Commonwealth Edison, the electric utility serving Northern Illinois including the City of Chicago). Since 1995, UTT has operated the world's largest ice-based thermal storage system, providing chilled water for air conditioning to downtown office buildings and hotels totaling more than 2.3 million m^2 of space in the city of Chicago.

UTT was also recently awarded a contract to heat and cool the new Midway Airport terminal and concourse in Chicago. For the Midway project, UTT will manufacture approximately 0.7 million kg of ice each night.

In 1996, UTT created a partnership with Boston Edison Company called Northwind Boston to supply DC in downtown Boston, Massachusetts. In 1997, Northwind Boston started construction of an ice-based central cooling plant in Boston's Back Bay. The ice plant will produce 1.4 million kg of ice at night. This plant will cool approximately 0.5 million m^2 of customer space in 1999. Plans for additional ice plants and expansion of the DC system in Boston are under development.

In 1997, Houston Industries Incorporated (parent of Houston Lighting and Power Co.) formed a joint venture with UTT to develop a DC system in downtown Houston, Texas. The new company, Northwind Houston, a non-regulated affiliate of Houston Industries was created to provide customers with DC services. A DC facility located downtown will provide Houston with a strategic infrastructure improvement. Construction began in 1998.

Energy Pacific and Atlantic Thermal Systems Inc.

These partners are working to supply the energy needs of the DreamWorks SKG animation campus in Glendale, California. DreamWorks SKG was formed in 1994 by Steven Spielberg, Jeffrey Katzenberg and David Geffen to produce live-action motion pictures.

Energy Pacific is a non-regulated joint venture of Pacific Enterprises, the Los Angeles-based parent of Southern California Gas Co. and Enova Corp., based in San Diego, the parent of San Diego Gas & Electric. ATS is one of a group of non-regulated energy services businesses held by Atlantic Energy Enterprises, a subsidiary of Atlantic Energy Inc., the parent of Atlantic Energy Electricity Co.

In a bundled package that is called integrated energy management services, Atlantic/Pacific will provide to DreamWorks the total energy supply on a cost-per-square foot basis. The billing amount will be based on the company's actual cost of constructing the central plant and other energy systems and will include a fixed component to cover the company's capital recovery and a variable component to cover its operating and maintenance costs. Atlantic/Pacific is also developing a DE system at the Venetian Hotel in Las

Vegas. ATS has already constructed two DE plants in Atlantic City, New Jersey: Bayside Plant for the 200,000 m² Convention Center and Midtown Plant for the Casino Areas.

The Public Service Company of Colorado

This venerable company, which supplied steam to downtown Denver since 1880 and DC from 1898 to 1918, started construction of a district chilled water system in downtown Denver in 1998. The ice-based plant will initially supply 7,000 tons of chilled water. When fully developed, the plant will provide approximately 25,000 tons of chilled water and can handle about one-third of downtown Denver's cooling needs. Public Service Co.'s investment will total more than $40 million over the next five years to complete the DC system. DC will help customers reduce their need for peak power which helps them better manage energy costs.

CTG Resources, Inc.

CTG is the holding company of Connecticut Natural Gas Corp. (CNG) and its unregulated subsidiary, The Energy Network (TEN); which through its DH/DC subsidiary, the Hartford Steam Co., provides 36,000 tons of cooling and heating to large customers in the downtown and capital area of Hartford, Connecticut. The Hartford Steam Co. has been providing DH/DC service for over 35 years. CTG is engaged in a number of new energy-supply businesses through TEN.

The Thirteen Largest District Heating Systems in the World[56]

Country	Scheme	Capacity MW	Annual Energy Production	
			PJ	GWh
1. Russia	St. Petersburg	17,440	237	66,000
2. Russia	Moscow	n.a.	150	42,000
3. Ukraine	Kyiv	13,000	133	37,000
4. Poland	Warsaw	n.a.	90	25,000
5. Czech Republic	Prague	n.a.	54	15,000
6. Belarus	Minsk	7,900	45	12,500
7. Germany	Berlin	6,056	40	11,000
8. Romania	Bucharest	n.a.	37	10,200
9. Republic of Korea	Seoul	n.a.	36	10,000
10. Poland	Krakow	3,770	n.a.	n.a.
11. United States	New York	3,516	30	8,400
12. Denmark	Copenhagen	n.a.	26	7,000
13. Finland	Helsinki	2,500	23	6,500

---

[56] Defining the largest schemes is problematic because the statistics concerning DH supply in one city may include one or more schemes. Also, there are no comprehensive statistics available on all the city-wide systems in the large countries of Russia and Ukraine, for example. The table has been prepared based on annual energy production in the later half of the 1990's. Different data sources have been used and as some schemes are growing (e.g. Seoul), the order may not be fully correct and should be viewed as mainly indicative.

Extent and Key Characteristics of Central and Eastern European and Former Soviet Union Countries

Russia

Today, Russia has the largest DH networks in the world. The total length of the networks (two pipe system) is estimated to be over 260,000 kilometers. Approximately 70% of all the residences in the country of about 150 million inhabitants are connected to the DH networks. Water systems comprise over 90% of the total, the remaining being steam systems.[57]

Latvia

DH systems are widespread in Latvia and operated in all major towns and even a major share of countryside towns and villages. Most of the DH schemes have been constructed between 1960 and 1990; a considerable part of the systems is at the end of its technical service time. 70% of all households are connected to the DH networks. In the capital city, Riga, 83% of the population was connected to DH in 1996.[58] The DH system of Riga is supplied by two CHP plants and various HOBs, but the number of HOBs has been rapidly decreasing.

Ukraine

In Ukraine, with a population of about 50 million, DH has reached a 65% market share of the heat market. The total annual heat consumption is 246 TWh. The main fuel is natural gas (about 70%) but also oil, coal and other fuels are used. It is very common that large factories supply heat and domestic hot water to local residential areas. CHP and DH face increasing competition from gas and electricity companies.

Lithuania

In Lithuania, DH was introduced in the late 1950's. About 59% of the households in the whole country and 85% in the capital city, Vilnius, are connected to the DH networks. The possibilities of increased CHP production in Lithuania are limited because there is no demand for the electricity output as long as there is the very cheap and abundant electricity available from the Ignalina nuclear power plant.[59] After the nuclear power plant is closed, the situation is likely to change.

Estonia

The market share of DH was 52% of whole heating market in 1996. In large towns, the market share is close to 90%. The main competitors are individual oil heating, electric heating and lately also local gas boilers.

Almost 80% of DH is produced in HOBs with capacity under 1 MW. The main fuels are natural gas (32%), heavy fuel oil (27%) and oil shale (14%). The share of natural gas is increasing. The relatively low share of CHP is explained by the low cost of electricity, but this is increasing recently. In CHP plants, the main fuel is oil shale (93%) followed by natural gas and oil.

---

[57] Koljonen, 1998.
[58] Koljonen, 1998.
[59] UNDP and ESMAP, 1998.

Poland

The market share of DH is 34% in the housing sector and 52% for the whole building stock. The main competitors of DH in the housing sector are coal and coke (60%) and oil and gas (6%).

The number of CHP plants was 265 in 1993; 57 of the industrial CHPs provide heat to the DH networks.[60] 55% of DH was provided by CHP plants in 1995. The main fuel used for DH production was coal with an 88% share in 1995.

Over 313 DH companies operated in Poland in 1993. Many of the DH companies purchase heat from an industrial supplier and only one-third own their heating plants. The majority of the DH companies are owned and controlled by municipalities, although small DH plants are often owned by housing co-operatives.[61]

Belarus

The market share of DH in the household heating sector is 50%. Most of the DH (78%) is produced in CHPs. In rural households, firewood is the main heating fuel.

The number of CHP plants is 22 with a total heat production capacity of 10 176 MW$_t$; the number of HOBs is 5,000 with a total heat production capacity of about the same. In addition, industrial boilers provide heat to the DH networks.

Slovak Republic

The market share of DH in the household heating sector is 40%. Most of these are blocks of flats in larger cities. The main fuels for DH production are natural gas (53%) and coal (28%).

Czech Republic

DH exists in most of the main towns in the country and one of the ten largest schemes in the world is installed in Prague. Today, the market share of DH is 33% in the residential sector. The main competitors are individual gas heating (27%), solid fuels (23%) and electric heating (9%). The government has favored individual gas heating. However, competition by gas is weakening, as price deregulation processes are expected to increase the price of competing heating forms more than that of DH.

Coal, mined in the Czech Republic, is the most important fuel for DH and CHP production. However, there is a trend towards increased use of natural gas. 25% of DH is produced in CHP plants.

Romania

Today, the market share of DH is 31% in the residential sector. In urban areas, the market share of DH is higher, 57%. The main competitors in heating are coal, oil and gas stoves (61%). The competition is expected to become more intense in the future.

In 1995, gas was the most important fuel in DH production with a share of 50% followed by oil (33%), coal (9%) and other sources (8%). 61% of DH is produced in CHP plants.

---

[60] Koljonen, 1998.
[61] Koljonen, 1998.

Bulgaria

The market share of DH is 19% in the residential sector. The largest scheme is in Sofia where almost 65% of the households are connected to DH. The main competitors in heating are coal and briquettes (45%), electric heating (21%) and oil (9%). Though electric heating has been 2-3 times more expensive than DH, due to the lack of high-grade indigenous fuels, it has become relatively significant.

The share of co-generated heat is very high at 82%. The main fuels in DH production are natural gas (66%), oil (26%) and coal (7%).

Hungary

The market share of DH is 17% in the residential sector representing some 650,000 households. The share of co-generated heat was 25% in 1996 and 48% in 1997. A couple of new CHPs were commissioned at the time but the rapid increase in CHP share may be partly a statistical error. The economic and legislative situations have hampered the introduction of additional CHP. The main fuels in DH production are natural gas (55%), oil (21%) and coal (20%). Other fuel sources under investigation are waste incineration, wood chips and geothermal energy.

Croatia

DH in concentrated in Osijek and Zagreb where the market share of DH is 25% and all production takes place in CHPs. There are some boilers in operation in other towns.

Yugoslavia

The market share of DH is 13%. DH exists only in cities and individual heating has been the most common form of heating due to the moderate climate. In 1996, the number of CHPs was 17 and boiler plants 196. Natural gas and coal are the most common fuels for heat production.

Slovenia

The market share of DH is 9% in the whole building stock. The main competitors are oil heating (35%) and individual gas heating (35%). The share of natural gas heating has been increasing due to the construction of new networks and low connection prices.

The share of co-generated heat is 64%. The main fuels in DH production are brown coal (46%), natural gas (29%) and lignite (18%).

Armenia

District heating can be found in Yerevan and in 54 counties and small cities. DH is produced in three thermal power plants and a number of small heating boiler houses. The total length of DH networks was about 900 kilometers in 1994. In 1997, heat production was about 3,700 TJ and heat consumption in the household and service sectors was about 1,700 TJ. Heat production has decreased since 1992, when the numbers were 11,200 and 5,900 TJ, respectively. Heat production has decreased from 1989 to about one-sixth of its former value.

The dramatic decrease in the municipal heat and hot water supply during the last five years has resulted in several negative ecological impacts. First, the absence of municipal heat and hot water supply caused the vast portion of the population to use electricity for heating purposes, which is an inefficient way to use energy resources. This situation became a heavy burden on the Armenian energy sector in general and caused additional energy losses and greenhouse gas emissions.

Another consequence of the DH network collapse is a massive harvesting of trees for heating and cooking purposes. According to the data of National Greenhouse Gas Inventory in 1994, the use of fuel wood in 1994 exceeded the level of 1990 by eight-fold. The massive forest cutting caused a disturbance of the ecological balance acquired during many centuries, as well as created a threat for new land territories.

Georgia

Before 1993, the heat and hot water supply in Georgia was based 99% on imported fuels – natural gas and oil (mazut). Heat was produced mainly in centralized DH plants with total thermal capacity of 4,295 MW in Tbilisi and 2,093 MW in other cities (Kutaisi 264 MW; Batumi 238 MW; Gori 104 MW; and Rustavi 289 MW).

In 1993, the heat and hot water supply and the gas supply for cooking were stopped practically all over the country, because the Government could not carry the rapidly rising costs of the imported fuels anymore. As a result, households started to use whatever means to survive, using kerosene, propane, wood, coal or electricity for their heating and cooking needs. During the last years, the only working plant in Tbilisi has been the CHP plant in the center of Tbilisi (18 MWe, 40 MW thermal), using natural gas and mazut as a fuel, and supplying heat mainly to the government buildings. In 1997, heat consumption was about 20,000 TJ mainly in the household and service sectors.

The development of the Tbilisi central heating system started in 1960. In 1989, about 80% of the heat was supplied by a total of 45 central heating stations operated by the municipal DH company, 6% from the Tbilisi thermal power station operated by the state power company, 5% from industrial boilers and 9% from autonomous boiler facilities. The centralized heating system supplied heat to 83% and hot water to 75% of the city population, serving about 340,000 households in 7,500 residential buildings and about 1,000 public and administrative buildings. The total length of the DH network in Tbilisi is 1,166 kilometers. The external and internal corrosion of the pipelines is estimated to be extensive.

Recently, about 30 new heating systems of the average size of 250-300 kW have been installed in schools and hospitals. In addition, some buildings are provided with the heat from six re-opened geothermal wells in the Lisi-Sees area. The total capacity of this geothermal reservoir has been estimated at 20-25 MW, which could serve approximately 3,000 households. More significant geothermal resources with temperatures between 90 and 100°C have been identified. Use of these reservoirs could open up completely new opportunities for heat and hot water supply for Tbilisi, as well as for the production of electric power.

Kazakhstan

The DH system in the capital city, Almaty, covers about 70% of the building volume in Almaty. The construction of the DH network in Almaty began in 1961. The system is supplied by two CHP plants and three peak and reserve boiler plants. A third CHP plant supplies DH to a separate DH system outside the city. Total installed heat capacity is 2,245 at the two CHP plants. The fuels used are coal, mazut and gas. The DH system in Almaty is unique. Consumers are connected to the network by open connection, and domestic hot water is taken directly from the DH system. One CHP plant produces only hot make-up water to replace the use of domestic hot water; water is not returned to the plant.

Kyrgyz Republic

The DH system in the capital city, Bishkek, is supplied with DH from the Bishkek Thermal Electric Station (TES) and boilers. The construction of TES began in 1958, and currently, its electric capacity is 609 MW_e and 1,817 Gcal/h of heat. In addition, there are twenty-four boilers and three natural gas-fired hot water peaking boilers. The total steam output of the boilers is 4,250 Gcal/h. Today, the length of the network is about 400 kilometers. In the city of Osh, the DH system is served by another CHP plant. The design fuel for the plant was natural gas but the plant uses currently mazut.

In 1996 the total heat demand about 3,200 GWh was composed of 67% residential, 10% industrial, 3% agricultural and 20% other consumption. In Bishkek, the total number of consumers was 3,411 and the total heating volume was about 31 million m^3.

The History and Status of Deregulation of the Gas and Electricity Markets in the United States

Gas Deregulation

Natural gas has long been regarded as another "natural monopoly," with local distribution systems coming under the regulation of state public service commissions and interstate production and transport overseen by the Federal Energy Regulatory Commission (FERC) and its predecessors. Until the "energy crisis" of the early 1970's, the wholesale price of gas was tightly regulated by the FERC. The regulatory practices led to a decline in investment in new wells and fears of significant shortages.

In response, the price of gas from new wells was gradually deregulated, and the prices paid by industry and utilities rose rapidly in the mid-1970's. Consumers responded with conservation measures and fuel switching which led to a significant decline in gas consumption. Gas prices peaked in the early 1980's and have declined steadily since then, leading to steady increases in consumption by industry and other consumers.

The most significant step in gas market development affecting DE systems was FERC Order 636 promulgated in 1993. This order required interstate gas pipeline companies to separate or "unbundle" their sales and transportation services. Prior to this, the owners of the pipelines had been able to dominate the wholesale market through control of the delivery mechanism. With full freedom to purchase bulk gas from any seller and be assured of cost-based pricing for pipeline use, industrial and utility customers have seen their deliveries through this "Sales for the Account of Others" mechanism rise from 35-40% of total gas purchases in 1986 to 75% in 1994. This flexibility ensures that any DE system near a suitable gas transmission corridor (and this now includes most populated areas of the United States) has the option to create whatever long or short-term contracts for gas supply they will find most useful. This development has improved the competitiveness of DE operations relative to other energy options.

Electric Utility Deregulation

In response to the pressures resulting from the "energy crisis" of 1973, the obvious benefits of co-generation and power production from renewable sources and the reluctance of the utilities to develop these sources or to integrate their electric systems with others who were developing them, Congress passed the Public Utilities Regulatory Policy Act of 1978 ("PURPA").

This law required utilities to interface with independent power producers, to provide them with back-up power, to purchase their excess power and to do so at rates reflecting the "full avoided costs." That is, the utility had to pay an outside provider for electric power at a rate that represented the full savings enjoyed by the utility because it did not have to supply that power, including long-term capital savings where appropriate. The rate specifically had nothing to do with the cost to the provider of generating the power, and this was done with the intent of fostering an independent power generation industry. To qualify for this rate, the provider was required to either produce the power from a renewable source (including small hydropower) or from a co-generation system meeting minimal standards for the amount of otherwise reject heat used. Under follow-up legislation, the utilities were also required to purchase power from any outside generator under less stringent price regulation.

The result was dramatic growth in non-utility generators (NUGs) and growth of the independent generation market from essentially nothing in 1977 to about 7% of the United States' generation in 1995. Building on this success, policy makers moved rapidly to a maximally deregulated model for the electric power industry. Requiring action at both Federal and state levels, the United States' power industry is adopting, or being forced to adopt, this model at the current time.

Under this system, the utilities sell off all of their assets except the distribution system, which they continue to operate as a regulated public utility with a natural monopoly. Generation stations are purchased and operated by independent companies, which can be anything from local one-unit enterprises to international conglomerates. Power is shipped from generators to the distribution utilities or directly to large customers by the transmission system, which is operated by an "Independent System Operator" (ISO) which is being formed from the institutional remnants of the earlier power pool operators. A key aspect of the new vision is that any consumer can arrange to purchase electrical energy from any producer and can arrange for its delivery with the payment of reasonable and regulated charges for the use of transmission and distribution facilities. Many complex arrangements have been settled or are now being worked out for the allocation of scarce transmission capability, assurance of adequate generation capacity, degree of support for past investments ("stranded assets") and many other significant issues.

Because the regulation of electric utilities was carried out by the individual states, consistently with FERC oversight and regional reliability councils, the deregulation process is being carried out on a state-by-state basis also. Table 3-1 below indicates how far along each state is; broadly, the Northeast and Southwest are most advanced, while the Northcentral and Southeastern states are moving more slowly.

TABLE 1
Electric Utility Restructuring Status

Level of Deregulation	States at that Level
Legislative/Regulatory Plan Adopted	AZ, CA, CT, IL, MA, MD, ME, MI, MT, NH, NJ, NV, NY, OK, PA, RI, VA
Pilot Program Approved	ID, NM, OR, WA
Plan or Pilot Program Pending	AK, DE, IA, KS, KY, NC, OH, SC, TX, VT, WI, WV
Legislative/Regulatory Investigations Ongoing	AL, CO, FL, GA, IN, LA, MN, MS, MO, NB, ND, SD, TN, UT, WY
Preliminary Activity Only	AK, HI

Technical Statistics of European District Energy Systems[1]

	1	2	3	4	5	6	7	8	9	10	11
	No. of under-takings	No. of CHP stations	No. of heating stations	Total length of pipelines	Max. DH output capacity	Subscribed demand	Load density	CHP heat output % of total DH energy	Heat taken by consumers	Maximum DC output capacity	DC consumed
				km	MW	MW	MW/km	%	TJ	MW	TJ
Austria	39 (40)	21 (28)	40 (50)	2 047 (2 409)	5 300 (5 900)	5 300 (5 500)	2.6 (2.3)	60 (65)	32 292 (35 802)	0 (n.a.)	0 (n.a.)
Belarus (1996)	n.a	22	1 437	5 000	36 210	n.a.	n.a.	10 176	n.a.	-	-
Bulgaria	-	28	30	2 600	8 800	8 960	3.4	52	2 418	-	-
Croatia	(1)	(3)	(4)	(258)	(1 916)	(921)	3.6	(n.a.)	(8 522)	-	-
Czech Republic	1 850 (1 928)	63 (163?)	1 355 (974)	9 600? (2 501?)	48 885 (39 372)	n.a. (n.a.)	n.a. (n.a.)	33 (25)	205 908 (n.a.)	- (0)	- (0)
Denmark	n.a. (425)	240 (250)	740 (750)	21 000 (22 000)	15 200 (15 200)	21 300 (22 400)	1.0 (1.0)	46 (61)	n.a. (n.a.)	- 0	- 0
Estonia (1996)	80	8	140	2 030	10 774	10 774	5.3	28	42 133	0	0
Finland	127 (127)	74 (78)	901 (895)	7 400 (7 880)	16 820 (17 470)	12 800 (13 300)	1.7 (1.7)	32 (79)	89 440 (94 820)	0 (n.a.)	0 (n.a.)
France	379	41	545	2 902	20 519	19 328	6.7	15	85 300	290 (290)	1 172 (1 169)
Germany	233	479	1 253	16 343	47 900	55 659	3.4	57	n.a.	57	-
Greece	(2)	(4)	(1)	(135)	(140)	(213)	1.6	(0)	(935)	(0)	(0)
Hungary (1996)	175	44	281	1 980	17 800	n.a.	n.a.	48	67 200	-	-
Iceland	31 (31)	1 (1)	-	2 784 (2 970)	1 499 (1 289)	n.a	n.a.	8 (10)	n.a. (1 583)	0 (0)	0 (0)
Italy	27 (27)	27 (28)	24 (8)	949? (762?)	2 462 (2 624)	2 114 (2 636)	2.2 (3.5)	45 (75)	8 658 (9 723)	52 (55)	117 (126)
Lithuania	(21)	(15)	(3 208)	(2 847)	(34 479)	(n.a.)	(n.a.)	(n.a)	(55 439)	(0)	(0)
Netherlands	17 (13)	35 (37)	110 (124)	2 403 (2 415)	3 979 (4 310)	3 900 (4 335)	1.6 (1.8)	60 (90)	12 930 (17 716)	- (6)	- (76)
Norway	17 (21)	2 (2)	25 (39)	303 (320)	750 (900)	635 (800)	2.1 (2.6)	8 (11)	4 159 (4 568)	8 (8)	9 (40)
Poland	720	250	6 940	14 000	57 210	51 500	3.7	47	361 895	-	-
Romania	-	118	1 745	16 510	48 206	n.a.	n.a.	57	400 880	-	-
Russia	163	123			1 445 000	n.a.	n.a.	13	-	-	-
Slovenia	27	3	94	481	1 797	1 910	4.0	27	8 434	0	0
Sweden	152 (162)	32 (34)	846 (865)	9 490 (9 964)	28 050 (28 200)	21 632 (25 593)	2.3 (2.6)	26 (28)	144 940 (148 129)	46 (137)	108 (504)
Switzerland	26	32	33	698	2 073	2 421	3.5	27	12 175	-	-
Ukraine	(13 507)	(274)	(30 754)	(44 931)	(195 856)	(n.a.)	(n.a.)	(n.a.)	(883 968)	(0)	(0)
United Kingdom	52	52	0?	n.a.	645	193	n.a.	5	5 850	-	-

Note: [1]Disregarding legal forms and ownership considerations, heat supply undertakings are all such undertakings and works which supply consumers with heat in the form of steam or water (public heat supply). [2] In a power and heat supply station electrical energy as well as DH is produced (CHP production). [3] A plant from which a supply of heat only is given. [4] The route length of a pipeline system is the simple length from the point of production to the point of delivery measured according to the street plan. The route length includes the house service connections. [5] The maximum heat output capacity is the maximum heat output, as limited by the weakest part of the plant, that can be sent out by the heat supplier. Both the plants owned by the DH undertakings as well as the plants owned by outside suppliers are included. [6] The arithmetic sum of capacities forming the supply contracts, actual or estimated, at the end of the reporting season. [7] Subscribed demand per pipeline length. [8] Total heat taken by consumers, measured at the supply point. [9] The maximum DC output capacity is the maximum DC output that can be sent out by the heat supplier. [10] Quantity of DC taken by the consumers, measured at the supply point.

___

[1] Data are from the year 1995 and data in parentheses are from the year 1997. Question marks display questionable data. Euroheat & Power, 1997 and 1999.

Largest District Cooling Systems in the United States

No.	Location	Owner	Capacity		Primary Energy	Type of Chillers
			Tons	MW		
1	Chicago, IL	Unicom/Northwind Chicago	86,000	302	Electricity	Electric-Driven
2	World Trade Center New York City, NY	Port Authority	50,000	176	Electricity and Gas	Electric and Steam-Driven
3	Houston, TX	Texas Medical Center	48,000	169	Electricity and Gas	Electric and Steam-Driven
4	Washington, DC	The Pentagon, U.S. Dept. of Defense	37,500	132	Electricity	Electric-Driven
5	Harvard University Boston, MA	Co-generation Management Company	36,000	127	Electricity and Gas	Electric and Steam-Driven
6	Indianapolis, IN	Mid-America Energy Resources	32,500	114	Electricity, Coal, Refuse	Steam and Electric-Driven
7	Hartford, CT	Connecticut Natural Gas/CTG Resources	32,000	113	Electricity and Gas	Electric and Steam-Driven
8	Minneapolis, MN	NRG	30,000	105	Electricity and Gas	Electric and Steam-Driven
9	Columbia, MO	Univ. of Missouri, Columbia	28,000	98	Electricity and Gas	Electric and Steam-Driven
10	Century City, CA	Sempra Energy Central Plants	27,000	95	Electricity and Gas	Electric and Steam-Driven
11	Lincoln, NB	University of Nebraska	25,200	89	Electricity and Gas	Electric and Steam-Driven
12	Nashville, TN	Nashville Thermal Transfer Corp.	24,000	84	Refuse	Steam-Turbine Driven
13	Chapel Hill, NC	Univ. of NC at Chapel Hill	23,400	82	Electricity and Gas	Electric and Steam-Driven
14	Tulsa, OK	Trigen Oklahoma	21,700	76	Electricity and Gas	Electric and Steam-Driven
15	Iowa City, IA	Univ. of Iowa	19,700	69	Electricity and Gas	Electric and Steam-Driven

District Heating and Energy Efficiency Projects Financed by the World Bank Group

Project	Year	Credit/Loan Amount (US$ million)	Total Cost (US$ million)
Projects Under Implementation:			
Poland Heat Supply Restructuring and Conservation Project	1991	$203	$739
China Beijing Environmental Project	1991	125	299
Estonia District Heating Rehabilitation Project	1994	39	65
Poland Coal-to-Gas Conversion Project (GEF)	1994	-	48
Poland Katowice Heat Supply Project	1995	45	93
Latvia Jelgava District Heating Rehabilitation Project	1995	14	18
Bosnia Emergency District Heating Reconstruction	1996	20	45
Kyrgyz Republic Power and District Heat Project	1996	20	98
Slovenia Environment Project	1996	24	38
Russia Energy Efficiency Project	1996	70	88
Lithuania Energy Efficiency/Housing Project	1996	10	21
Russia Enterprise Housing Divestiture Project	1996	300	551
Bulgaria District Heating Pilot Project (amendment to Water Companies Restructuring and Modernization Project)	1997	10	13
Czech Republic Kladno Co-generation Project	1997	125	375
Lithuania Klaipeda Geothermal Demonstration Project (GEF)	1997	6	18
China Shandong Environment Project (DH component)	1997	44	89
Russia Severstal Heat and Power Project	1998	67	102
Ukraine Kiev District Heating Improvement Project	1998	200	309
Czech Republic Kyjov Waste Heat Utilization Project (GEF)	1998	-	25
Ukraine Kiev Public Buildings Energy Efficiency Project	2000	18	30
Total		**1,340**	**3,064**

District Heating and Energy Efficiency Projects Financed by the World Bank Group (cont'd.)

Projects Under Preparation:

Poland Podhale Geothermal District Heating and Environment Project (GEF)

Ukraine Sevastopol Heat Supply Improvement Project

Russia Municipal Heating Project

Latvia Riga District Heating Rehabilitation Project

Lithuania Vilnius District Heating Rehabilitation Project

Bulgaria District Heating Project

Hungary Szombathely CHP/Biomass Project (GEF)

Hungary Szekesfehervar CHP/Biomass Project (GEF)

Croatia District Heating Rehabilitation Project

Poland Krakow Energy Efficiency Project (GEF)

Slovak Republic Industrial Co-generation Project (GEF)

Slovak Republic Kosice Geothermal District Heating Project (GEF)

Slovenia Carbon-Based Loan Expansion Project (GEF)

Results of the Poland District Heating Projects

The Polish energy sector was brought to the forefront of the World Bank-Poland dialogue in the late 1980s. At that time, the energy sector was characterized by its high energy intensity, over-centralization resulting in vast institutional inefficiencies and operational wastage, dependence on the state budget for investment and price subsidies, and the absence of market-related price signals (except for coal exports). In addition, it was the country's major source of air pollution resulting primarily from burning coal for power generation and heating. In 1990, the Government initiated the implementation of a reform program to improve the economic efficiency of the sector. This program involved the restructuring of energy enterprises and the reduction of the sector's fiscal impact. The latter was to be achieved by eliminating budget subsidies to energy enterprises and cross-subsidies between industrial and household consumers and by gradually phasing out subsidies to households. In addition, the program provided for energy prices to reach economic level either by decontrolling prices completely or by setting up a regulatory system with appropriate pricing rules.

The district heating (DH) sector recognized these inefficiency problems. For a considerable period of time, district heating enterprises (DHEs) suffered from a lack of funds to effectively operate, maintain and renew their infrastructure. This resulted in major technical problems including, inter alia, obsolete technology, serious corrosion problems caused by poor water quality and water leakage, lack of insulation resulting in major heat losses. The decentralization of ownership of DHEs to local municipalities in early 1990s combined with the elimination of state-subsidies for investments and the lack of long-term financing for infrastructure further exacerbated their financial situation. The local municipalities did not have the technical or financial resources to address these problems and the backlog of investments required both to preserve the enterprise's operating capability and to provide competitive, reliable and affordable DH services to the customers.

The Polish DH sector merited the World Bank support as the actions to increase the sector efficiencies would yield immediate large benefits in terms of coal savings and lower pollution. The World Bank financed DH projects in five cities (Gdansk, Gdynia, Krakow and Warsaw in 1991, and Katowice in 1994). Each project addressed aspects specific to the borrowing entity and was designed to improve organizational, technical, financial and environmental situations. The project also addressed other broader aspects related to energy pricing policies. In all five cities, the investment measures financed by the World Bank were identified in the context of master plan studies and were designed to extend the life of the district heating assets, improve the system operation, secure energy savings and reduce environmental pollution. The projects, commenced in 1991, are now completed. The results as described below are highly satisfactory and surpass expectations.

The total completion cost of the four projects is US$470 million and was financed by World Bank loans of US$165 million and the enterprises' internal funds of US$305 million. Table below shows the completion cost and financing plan for each city. [62/]

US$ million	Gdansk	Gdynia	Krakow	Warsaw	Total
IBRD Loans	40	25	25	75	165
DHE Fund	35	21	47	202	305
Total Costs	75	46	72	277	470

[62/]   DHE in Katowice received a World Bank loan of US$45 million and contributed an equivalent amount from its own funds for a similar heat supply project. The project in Katowice is expected to be completed in 2001.

The projects have achieved their objectives and the benefits can be summarized as follows:

- **Fiscal:** Investment subsidies were eliminated, and household subsidies phased out gradually from a nationwide average of 78% of the heating bill in 1991 to zero in 1998.

- **Consumers:** The efficiency gains resulting from the Government's energy pricing policy and achieved by both the DHEs and the combined heat and power plants greatly benefited the DH customers through a 55% lower price for heating one square meter (m^2) of floor area (from 54.5 PLN/m^2 in 1991 to 24.5 PLN/ m^2 in 1999, at 1999 prices).

- **Efficiency Gains:** 22% energy savings in the four cities was achieved and valued at US$55 million per year. The Bank-financed investments were focused on sustaining the least-cost heat solution, with significant benefits for the environment.

- **Financial:** Despite the reduction in their profit margins due to real tariff decreases, the enterprises were able to generate cash internally of 62% of capital investments, exceeding the minimum level of 30% required under the financial covenants of the World Bank loans.

- **Technical:** Control of the district heating systems was automated and changed from production control to demand control, thus giving the customer the possibility to regulate their heat consumption.

- **Environmental:** The citizens benefited from improved air quality through reduction in gaseous and dust emissions. The Krakow heating company was officially eliminated from the list of heavy air polluters, thanks to the boiler elimination and coal-to-gas conversion program. The Polish economy also reduced its contribution to greenhouse gases (mainly from carbon dioxide emissions).

- **Capacity Building of District Heating Staff and Management:** Comprehensive training programs were implemented, which extended the expertise of already technically qualified staff in other areas of economics, finance, international accounting, quality assurance, marketing and customer relations, and in modern district heating management and operations techniques and technologies.

- **Staff Productivity:** The productivity increased: staff were reduced by 32% and number of customers increased by 12%.

- **Institutional Development:** The enterprises were transformed from state production oriented organizations into market oriented enterprises:
  - Marketing and public relations departments were established
  - Expertise in international competitive procurement and trade was developed
  - Collection departments were reinforced and receivables reduced from 90 to 60 days on average
  - Financial accounts were computerized
  - Management and cost accounting systems were developed
  - Customer energy audits were undertaken
  - Quality assurance programs were (or are being) implemented
  - All customer are now metered.

- **International Cooperation:** The project promoted cooperation between the four district heating companies and other companies in Poland and abroad. Staff from the Polish companies have provided assistance to project implementation units of other World Bank-funded projects, in the areas of DH master plans, project design, preparation, procurement, disbursement and implementation. Similar assistance was provided to district heating projects and programs funded by the US Agency for International Development, the US Department of Energy, and the European Bank for Reconstruction and Development.

- **Local Industry:** The project in the four cities stimulated the local industry to become internationally competitive in the manufacturing and installation of modern district heating equipment. Major international equipment suppliers established manufacturing capability in Poland, and several of them formed joint-ventures with Polish partners. The share of the locally produced goods and installation services, financed by the World Bank and procured under international competitive bidding procedure, has increased from about 3% of the total value of goods, installation works and consultant services during 1992-94 to 50% during 1998-2000. This local manufacturing industry is available to supply the Polish DH market with annual capital investment estimated at US$350 million. This industry is also competing for ongoing Bank-financed projects in other countries.

Performance Indicators for the completed projects are presented below.

Indicators		Gdansk	Gdynia	Krakow	Warsaw	Total
Coal savings	('000 tons)	125	125	215	630	1,095
	% reduction	23%	25%	25%	21%	22%
Metering of sales	Before	14%	15%	16%	24%	21%
	After	100%	100%	100%	100%	100%
Reduction in water losses		26%	69%	48%	69%	65%
Reduction in number of staff		30%	33%	35%	30%	32%
Increase in productivity (TJ sales/employee)		11%	60%	21%	38%	34%
Reduction in air pollution (various emissions)		26-52%	30-42%	27-53%	25-42%	26-45%
Internal cash generation as % of investments		49%	52%	74%	64%	62%
Training (manweeks)		4,310	1,854	4,984	9,004	20,152
Technical assistance services (manweeks)		295	225	170	330	1,020

References

A. European

Delbès J. and Vadrot A. (1997): District Cooling Handbook. European marketing Group, District Heating and Cooling. With support from the EU THERMIE programme.

Ekono Energy Ltd (1997): Feasibility Study for Heat and Energy Efficiency Project; Kyiv, Ukraine. The World Bank.

Euroheat and Power (1999): Yearbook 1999.

Euroheat and Power (1997): Yearbook 1997. ISBN 3-8022-0513-8.

European Commission (1997): A Community strategy to promote combined heat and power (CHP) and to dismantle barriers to its development. Communication from the Commission to the Council and the European Parliament. Brussels 15.10.1998, COM(97) 514 final.

European Commission (1992). SEC (92)1411, final of 22 July 1992.

Government Statistical Service GSS (1997): Digest of United Kingdom Energy Statistics. Department of Trade and Industry. ISBN 011-515453-1

Green D. (1995): Successful Development of Community Heating and Combined Heat and Power. Combined Heat and Power Association, UK. Proceedings of the 27th Unichal Congress, Stockholm, Sweden, 12-14 June 1995.

Gunnarsdóttir M.J. (1998): Environmental Strategy in Nordic District Heating. In the proceedings of the Nordic-Baltic District Heating Symposium, 23-26 August 1998. Helsinki, Finland.

Hasenkopf (1995): Fernwärme in Europa unter Besonderer Berücksichtigung der Kraft-Wärme-Kopplung sowie ersten Ansätzen zum umweltgerechten Besteuerung von Energieträgern. Proceedings of the 27th Unichal Congress, Stockholm, Sweden, 12-14 June 1995.

Huovilainen R. and Koskelainen L. (1982): Kaukolämmitys (in Finnish). ISBN 951-763-209-6. Lappeenranta.

Koljonen T. (1998): District Heating Systems of MODIS Target Area. VTT Energy. Draft Report for Work Package 1, September 1998.

Ministry of Energy and Danish Energy Agency (1993): District Heating, Research and Technological Development in Denmark. ISBN 87-89072-71-5.

Power Economics: Restructuring District Heating in Central and Eastern Europe. Author Herkko Lehdonvirta, Ekono Energy Ltd. May 1998.

Swedish District Heating Association (1997): Dagsläge Fjärrkyla i Swerige, feb. 1997.

UNDP and ESMAP (1998): Increasing the Efficiency of Heating Systems in Central and Eastern Europe and the Former Soviet Union. The World Bank, Energy Unit.

US DOE (1999): Fossil Energy International at http://www.fe.doe.gow/int/, country reports from Eastern Europe.

Utility Europe: District Heating - Hot Prospects. June 1998.

Westin P. (1998): District Cooling in the World. Proceedings of the District Cooling Seminar, Stockholm, Sweden 25-26, 1998.

UNDP Project Document: Removing Barriers to Energy Efficiency in Municipal Heat and Hot Water Supply.

B. North American

"District Heating and Cooling in the United States: Prospects and Issues", National Research Council, National Academy Press, Washington DC 1985.

Federal Energy Regulatory Commission, 1984-1999.

"Annual Energy Review, 1995", U.S. Energy Information Administration, U.S. Government Printing Office, 1996.

"National Census for District Heating, Cooling and Co-generation", Oak Ridge National Laboratory (NTIS), 1993.

"District Heating and Cooling Systems for Communities Through Power Plant Retrofit Distribution Network", Public Service Electric and Gas Co. (NJ); National Technical Information Service, DOE/CS/20071-1, 1984.

"Scenarios of U.S. Carbon Reductions" (the "Five Labs Study") prepared by Lawrence Berkeley National Laboratory, Oak Ridge National Laboratory and Collaborators: ORNL/CON-44, September, 1997.

"Energy Innovations: A Prosperous Path to a Clean Environment", The Alliance to Save Energy (American Council for an Energy Efficient Economy, Natural Resources Defense Council, Tellus Institute and Union of Concerned Scientists), 1997.

Euroheat and Unichal Power Study Committee for Nomenclature and Statistics, 1995.

"A Community Strategy to Promote CHP and to Dismantle Barriers to its Development", The European Commission, Brussels COM (97), 514 Final, 15 November 1997.

Oliker, I., "District Energy Development in the United States", Proceedings of the 7th International Symposium on District Heating and Cooling, Nordic Energy Research Program, Lund, Sweden, May 18-20, 1999.

Oliker, I., "Fifteen Years of Experience with Combined Heat and Power in Jamestown, NY", NYSERDA Workshop, October 1999.

Oliker, I. "District Heating and Cooling/Co-generation" A Key to Sustainability", Proceeding of the 88th Annual Conference of the International District Heating Association, San Diego, CA, June 1997.

Oliker, I. "Converting Power Plants to District Energy Supply", Proceeding of the 87th Annual Conference of the International District Heating Assocation, June 1996.

Armor, A.F., Petrill, E. and Oliker, I, "EPRI Assessment of Power Plant Retrofits for District Heating and Cooling, Proceedings of the Heat Rate Improvement Conference, Baltimore, MD, May 3-5, 1994.

Oliker, I; Taranov, D; Belval, R; Irving, J; Maker, T. "Development of a District Heating System in Burlington Vermont", Proceeding of the 89th Annual Conference of the International District Energy Association, San Antonio, TX, June 13-16, 1998.

Oliker, I., Major, W., Drake, J., Wendland, R., "District Heating and Cooling Development in the United States", UNICHAL, Paris, June 8-10, 1993.

Oliker, I. and Major, W. "Impact of District Energy and Large Cool Storage on Demand Reduction, Electric Power Research Institute, 1992.

Distributors of World Bank Group Publications

Prices and credit terms vary from country to country. Consult your local distributor before placing an order.

ARGENTINA
World Publications SA
Av. Cordoba 1877
1120 Ciudad de Buenos Aires
Tel: (54 11) 4815-8156
Fax: (54 11) 4815-8156
E-mail: wpbooks@infovia.com.ar

AUSTRALIA, FIJI, PAPUA NEW GUINEA, SOLOMON ISLANDS, VANUATU, AND SAMOA
D.A. Information Services
648 Whitehorse Road
Mitcham 3132, Victoria
Tel: (61) 3 9210 7777
Fax: (61) 3 9210 7788
E-mail: service@dadirect.com.au
URL: http://www.dadirect.com.au

AUSTRIA
Gerold and Co.
Weihburggasse 26
A-1011 Wien
Tel: (43 1) 512-47-31-0
Fax: (43 1) 512-47-31-29
URL: http://www.gerold.co/at.online

BANGLADESH
Micro Industries Development
Assistance Society (MIDAS)
House 5, Road 16
Dhanmondi R/Area
Dhaka 1209
Tel: (880 2) 326427
Fax: (880 2) 811188

BELGIUM
Jean De Lannoy
Av. du Roi 202
1060 Brussels
Tel: (32 2) 538-5169
Fax: (32 2) 538-0841

BRAZIL
Publicacões Tecnicas Internacionais
Ltda.
Rua Peixoto Gomide, 209
01409 Sao Paulo, SP.
Tel: (55 11) 259-6644
Fax: (55 11) 258-6990
E-mail: postmaster@pti.uol.br
URL: http://www.uol.br

CANADA
Renouf Publishing Co. Ltd.
5369 Canotek Road
Ottawa, Ontario K1J 9J3
Tel: (613) 745-2665
Fax: (613) 745-7660
E-mail:
order.dept@renoufbooks.com
URL: http:// www.renoufbooks.com

CHINA
China Financial & Economic
Publishing House
8, Da Fo Si Dong Jie
Beijing
Tel: (86 10) 6401-7365
Fax: (86 10) 6401-7365

China Book Import Centre
P.O. Box 2825
Beijing

Chinese Corporation for Promotion
of Humanities
52, You Fang Hu Tong,
Xuan Nei Da Jie
Beijing
Tel: (86 10) 660 72 494
Fax: (86 10) 660 72 494

COLOMBIA
Infoenlace Ltda.
Carrera 6 No. 51-21
Apartado Aereo 34270
Santafé de Bogotá, D.C.
Tel: (57 1) 285-2798
Fax: (57 1) 285-2798

COTE D'IVOIRE
Center d'Edition et de Diffusion
Africaines (CEDA)
04 B.P. 541
Abidjan 04
Tel: (225) 24 6510; 24 6511
Fax: (225) 25 0567

CYPRUS
Center for Applied Research
Cyprus College
6, Diogenes Street, Engomi
P.O. Box 2006
Nicosia
Tel: (357 2) 59-0730
Fax: (357 2) 66-2051

CZECH REPUBLIC
USIS, NIS Prodejna
Havelkova 22
130 00 Prague 3
Tel: (420 2) 2423 1486
Fax: (420 2) 2423 1114
URL: http://www.nis.cz/

DENMARK
SamfundsLitteratur
Rosenoerns Allé 11
DK-1970 Frederiksberg C
Tel: (45 35) 351942
Fax: (45 35) 357822
URL: http://www.sl.cbs.dk

ECUADOR
Libri Mundi
Libreria Internacional
P.O. Box 17-01-3029
Juan Leon Mera 851
Quito
Tel: (593 2) 521-606; (593 2) 544-185
Fax: (593 2) 504-209
E-mail: librimu1@librimundi.com.ec
E-mail: librimu2@librimundi.com.ec

CODEU
Ruiz de Castilla 763, Edif. Expocolor
Primer piso, Of. #2
Quito
Tel/Fax: (593 2) 507-383; 253-091
E-mail: codeu@impsat.net.ec

EGYPT, ARAB REPUBLIC OF
Al Ahram Distribution Agency
Al Galaa Street
Cairo
Tel: (20 2) 578-6083
Fax: (20 2) 578-6833

The Middle East Observer
41, Sherif Street
Cairo
Tel: (20 2) 393-9732
Fax: (20 2) 393-9732

FINLAND
Akateeminen Kirjakauppa
P.O. Box 128
FIN-00101 Helsinki
Tel: (358 0) 121 4418
Fax: (358 0) 121-4435
E-mail: akatilaus@stockmann.fi
URL: http://www.akateeminen.com

FRANCE
Editions Eska; DBJ
48, rue Gay Lussac
75005 Paris
Tel: (33-1) 55-42-73-08
Fax: (33-1) 43-29-91-67

GERMANY
UNO-Verlag
Poppelsdorfer Allee 55
53115 Bonn
Tel: (49 228) 949020
Fax: (49 228) 217492
URL: http://www.uno-verlag.de
E-mail: unoverlag@aol.com

GHANA
Epp Books Services
P.O. Box 44
TUC
Accra
Tel: 223 21 778843
Fax: 223 21 779099

GREECE
Papasotiriou S.A.
35, Stournara Str.
106 82 Athens
Tel: (30 1) 364-1826
Fax: (30 1) 364-8254

HAITI
Culture Diffusion
5, Rue Capois
C.P. 257
Port-au-Prince
Tel: (509) 23 9260
Fax: (509) 23 4858

HONG KONG, CHINA; MACAO
Asia 2000 Ltd.
Sales & Circulation Department
302 Seabird House
22-28 Wyndham Street, Central
Hong Kong, China
Tel: (852) 2530-1409
Fax: (852) 2526-1107
E-mail: sales@asia2000.com.hk
URL: http://www.asia2000.com.hk

HUNGARY
Euro Info Service
Margitszgeti Europa Haz
H-1138 Budapest
Tel: (36 1) 350 80 24, 350 80 25
Fax: (36 1) 350 90 32
E-mail: euroinfo@mail.matav.hu

INDIA
Allied Publishers Ltd.
751 Mount Road
Madras - 600 002
Tel: (91 44) 852-3938
Fax: (91 44) 852-0649

INDONESIA
Pt. Indira Limited
Jalan Borobudur 20
P.O. Box 181
Jakarta 10320
Tel: (62 21) 390-4290
Fax: (62 21) 390-4289

IRAN
Ketab Sara Co. Publishers
Khaled Eslamboli Ave., 6th Street
Delafrooz Alley No. 8
P.O. Box 15745-733
Tehran 15117
Tel: (98 21) 8717819; 8716104
Fax: (98 21) 8712479
E-mail: ketab-sara@neda.net.ir

Kowkab Publishers
P.O. Box 19575-511
Tehran
Tel: (98 21) 258-3723
Fax: (98 21) 258-3723

IRELAND
Government Supplies Agency
Oifig an tSoláthair
4-5 Harcourt Road
Dublin 2
Tel: (353 1) 661-3111
Fax: (353 1) 475-2670

ISRAEL
Yozmot Literature Ltd.
P.O. Box 56055
3 Yohanan Hasandlar Street
Tel Aviv 61560
Tel: (972 3) 5285-397
Fax: (972 3) 5285-397

R.O.Y. International
PO Box 13056
Tel Aviv 61130
Tel: (972 3) 649 9469
Fax: (972 3) 648 6039
E-mail: royil@netvision.net.il
URL: http://www.royint.co.il

Palestinian Authority/Middle East
Index Information Services
P.O.B. 19502 Jerusalem
Tel: (972 2) 6271219
Fax: (972 2) 6271634

ITALY, LIBERIA
Licosa Commissionaria Sansoni SPA
Via Duca Di Calabria, 1/1
Casella Postale 552
50125 Firenze
Tel: (39 55) 645-415
Fax: (39 55) 641-257
E-mail: licosa@ftbcc.it
URL: http://www.ftbcc.it/licosa

JAMAICA
Ian Randle Publishers Ltd.
206 Old Hope Road, Kingston 6
Tel: 876-927-2085
Fax: 876-977-0243
E-mail: irpl@colis.com

JAPAN
Eastern Book Service
3-13 Hongo 3-chome, Bunkyo-ku
Tokyo 113
Tel: (81 3) 3818-0861
Fax: (81 3) 3818-0864
E-mail: orders@svt-ebs.co.jp
URL:
http://www.bekkoame.or.jp/~svt-ebs

KENYA
Africa Book Service (E.A.) Ltd.
Quaran House, Mfangano Street
P.O. Box 45245
Nairobi
Tel: (254 2) 223 641
Fax: (254 2) 330 272

Legacy Books
Loita House
Mezzanine 1
P.O. Box 68077
Nairobi
Tel: (254) 2-330853, 221426
Fax: (254) 2-330854, 561654
E-mail: Legacy@form-net.com

KOREA, REPUBLIC OF
Dayang Books Trading Co.
International Division
783-20, Pangba Bon-Dong,
Socho-ku
Seoul
Tel: (82 2) 536-9555
Fax: (82 2) 536-0025
E-mail: seamap@chollian.net

Eulyoo Publishing Co., Ltd.
46-1, Susong-Dong
Jongro-Gu
Seoul
Tel: (82 2) 734-3515
Fax: (82 2) 732-9154

LEBANON
Librairie du Liban
P.O. Box 11-9232
Beirut
Tel: (961 9) 217 944
Fax: (961 9) 217 434
E-mail: hsayegh@librairie-du-liban.com.lb
URL: http://www.librairie-du-liban.com.lb

MALAYSIA
University of Malaya Cooperative
Bookshop, Limited
P.O. Box 1127
Jalan Pantai Baru
59700 Kuala Lumpur
Tel: (60 3) 756-5000
Fax: (60 3) 755-4424
E-mail: umkoop@tm.net.my

MEXICO
INFOTEC
Av. San Fernando No. 37
Col. Toriello Guerra
14050 Mexico, D.F.
Tel: (52 5) 624-2800
Fax: (52 5) 624-2822
E-mail: infotec@rtn.net.mx
URL: http://rtn.net.mx

Mundi-Prensa Mexico S.A. de C.V.
c/Rio Panuco, 141-Colonia
Cuauhtemoc
06500 Mexico, D.F.
Tel: (52 5) 533-5658
Fax: (52 5) 514-6799

NEPAL
Everest Media International Services
(P.) Ltd.
GPO Box 5443
Kathmandu
Tel: (977 1) 416 026
Fax: (977 1) 224 431

NETHERLANDS
De Lindeboom/Internationale
Publicaties b.v.-
P.O. Box 202, 7480 AE Haaksbergen
Tel: (31 53) 574-0004
Fax: (31 53) 572-9296
E-mail: lindeboo@worldonline.nl
URL: http://www.worldonline.nl/~lin-deboo

NEW ZEALAND
EBSCO NZ Ltd.
Private Mail Bag 99914
New Market
Auckland
Tel: (64 9) 524-8119
Fax: (64 9) 524-8067

NIGERIA
University Press Limited
Three Crowns Building Jericho
Private Mail Bag 5095
Ibadan
Tel: (234 22) 41-1356
Fax: (234 22) 41-2056

PAKISTAN
Mirza Book Agency
65, Shahrah-e-Quaid-e-Azam
Lahore 54000
Tel: (92 42) 735 3601
Fax: (92 42) 576 3714

Oxford University Press
5 Bangalore Town
Sharae Faisal
PO Box 13033
Karachi-75350
Tel: (92 21) 446307
Fax: (92 21) 4547640
E-mail: ouppak@TheOffice.net

Pak Book Corporation
Aziz Chambers 21, Queen's Road
Lahore
Tel: (92 42) 636 3222; 636 0885
Fax: (92 42) 636 2328
E-mail: pbc@brain.net.pk

PERU
Editorial Desarrollo SA
Apartado 3824, Ica 242 OF. 106
Lima 1
Tel: (51 14) 285380
Fax: (51 14) 286628

PHILIPPINES
International Booksource Center Inc.
1127-A Antipolo St, Barangay,
Venezuela
Makati City
Tel: (63 2) 896 6501; 6505; 6507
Fax: (63 2) 896 1741

POLAND
International Publishing Service
Ul. Piekna 31/37
00-677 Warzawa
Tel: (48 2) 628-6089
Fax: (48 2) 621-7255
E-mail: books%ips@ikp.atm.com.pl
URL:
http://www.ipscg.waw.pl/ips/export

PORTUGAL
Livraria Portugal
Apartado 2681, Rua Do Carm
o 70-74
1200 Lisbon
Tel: (1) 347-4982
Fax: (1) 347-0264

ROMANIA
Compani De Librarii Bucuresti S.A.
Str. Lipscani no. 26, sector 3
Bucharest
Tel: (40 1) 313 9645
Fax: (40 1) 312 4000

RUSSIAN FEDERATION
Isdatelstvo <Ves Mir>
9a, Kolpachniy Pereulok
Moscow 101831
Tel: (7 095) 917 87 49
Fax: (7 095) 917 92 59
ozimarin@glasnet.ru

SINGAPORE; TAIWAN, CHINA MYANMAR; BRUNEI
Hemisphere Publication Services
41 Kallang Pudding Road #04-03
Golden Wheel Building
Singapore 349316
Tel: (65) 741-5166
Fax: (65) 742-9356
E-mail: ashgate@asianconnect.com

SLOVENIA
Gospodarski vestnik Publishing
Group
Dunajska cesta 5
1000 Ljubljana
Tel: (386 61) 133 83 47; 132 12 30
Fax: (386 61) 133 80 30
E-mail: repansekj@gvestnik.si

SOUTH AFRICA, BOTSWANA
For single titles:
Oxford University Press Southern
Africa
Vasco Boulevard, Goodwood
P.O. Box 12119, N1 City 7463
Cape Town
Tel: (27 21) 595 4400
Fax: (27 21) 595 4430
E-mail: oxford@oup.co.za

For subscription orders:
International Subscription Service
P.O. Box 41095
Craighall
Johannesburg 2024
Tel: (27 11) 880-1448
Fax: (27 11) 880-6248
E-mail: iss@is.co.za

SPAIN
Mundi-Prensa Libros, S.A.
Castello 37
28001 Madrid
Tel: (34 91) 4 363700
Fax: (34 91) 5 753998
E-mail: libreria@mundiprensa.es
URL: http://www.mundiprensa.com/

Mundi-Prensa Barcelona
Consell de Cent, 391
08009 Barcelona
Tel: (34 3) 488-3492
Fax: (34 3) 487-7659
E-mail: barcelona@mundiprensa.es

SRI LANKA, THE MALDIVES
Lake House Bookshop
100, Sir Chittampalam Gardiner
Mawatha
Colombo 2
Tel: (94 1) 32105
Fax: (94 1) 432104
E-mail: LHL@sri.lanka.net

SWEDEN
Wennergren-Williams AB
P. O. Box 1305
S-171 25 Solna
Tel: (46 8) 705-97-50
Fax: (46 8) 27-00-71
E-mail: mail@wwi.se

SWITZERLAND
Librairie Payot Service Institutionnel
C(tm)tes-de-Montbenon 30
1002 Lausanne
Tel: (41 21) 341-3229
Fax: (41 21) 341-3235

ADECO Van Diermen
EditionsTechniques
Ch. de Lacuez 41
CH1807 Blonay
Tel: (41 21) 943 2673
Fax: (41 21) 943 3605

THAILAND
Central Books Distribution
306 Silom Road
Bangkok 10500
Tel: (66 2) 2336930-9
Fax: (66 2) 237-8321

TRINIDAD & TOBAGO AND THE CARRIBBEAN
Systematics Studies Ltd.
St. Augustine Shopping Center
Eastern Main Road, St. Augustine
Trinidad & Tobago, West Indies
Tel: (868) 645-8466
Fax: (868) 645-8467
E-mail: tobe@trinidad.net

UGANDA
Gustro Ltd.
PO Box 9997, Madhvani Building
Plot 16/4 Jinja Rd.
Kampala
Tel: (256 41) 251 467
Fax: (256 41) 251 468
E-mail: gus@swiftuganda.com

UNITED KINGDOM
Microinfo Ltd.
P.O. Box 3, Omega Park, Alton,
Hampshire GU34 2PG
England
Tel: (44 1420) 86848
Fax: (44 1420) 89889
E-mail: wbank@microinfo.co.uk
URL: http://www.microinfo.co.uk

The Stationery Office
51 Nine Elms Lane
London SW8 5DR
Tel: (44 171) 873-8400
Fax: (44 171) 873-8242
URL: http://www.the-stationery-office.co.uk/

VENEZUELA
Tecni-Ciencia Libros, S.A.
Centro Cuidad Comercial Tamanco
Nivel C2, Caracas
Tel: (58 2) 959 5547; 5035; 0016
Fax: (58 2) 959 5636

ZAMBIA
University Bookshop, University of
Zambia
Great East Road Campus
P.O. Box 32379
Lusaka
Tel: (260 1) 252 576
Fax: (260 1) 253 952

ZIMBABWE
Academic and Baobab Books (Pvt.)
Ltd.
4 Conald Road, Graniteside
P.O. Box 567
Harare
Tel: 263 4 755035
Fax: 263 4 781913